T0248584

METAL IONS IN
BIOLOGICAL SYSTEMS

VOLUME 14

Inorganic Drugs in Deficiency and Disease

METAL IONS IN BIOLOGICAL SYSTEMS

Edited by

Helmut Sigel

Institute of Inorganic Chemistry
University of Basel
Basel, Switzerland

with the assistance of Astrid Sigel

VOLUME 14
Inorganic Drugs in Deficiency and Disease

CRC Press
Taylor & Francis Group
Boca Raton London New York

CRC Press is an imprint of the
Taylor & Francis Group, an **informa** business

First published 1982 by Marcel Dekker, Inc.

Published 2019 by CRC Press
Taylor & Francis Group
6000 Broken Sound Parkway NW, Suite 300
Boca Raton, FL 33487-2742

© 1982 by Taylor & Francis Group, LLC
CRC Press is an imprint of Taylor & Francis Group, an Informa business

First issued in paperback 2019

No claim to original U.S. Government works

ISBN 13: 978-0-367-45195-0 (pbk)
ISBN 13: 978-0-8247-1569-4 (hbk)
ISSN: 0161-5149

**Visit the Taylor & Francis Web site at
http://www.taylorandfrancis.com**

**and the CRC Press Web site at
http://www.crcpress.com**

Library of Congress Cataloging in Publication Data
Main entry under title:

Properties of copper.

    (Metal ions in biological systems, ISSN
0161-5149 ; v. 12)
    Includes bibliographical references and
indexes.
    1. Copper in the body. 2. Copper--
Physiological effect. I. Sigel, Helmut.
II. Sigel, Astrid. III. Series.
QP532.M47 vol. 12 [QP535.C9] 574.1'9214s 81-12587
ISBN 0-8247-1569-1 [599.01'9214]   AACR2

PREFACE TO THE SERIES

Recently, the importance of metal ions to the vital functions of living organisms, hence their health and well-being, has become increasingly apparent. As a result, the long-neglected field of "bioinorganic chemistry" is now developing at a rapid pace. The research centers on the synthesis, stability, formation, structure, and reactivity of biological metal ion-containing compounds of low and high molecular weight. The metabolism and transport of metal ions and their complexes is being studied, and new models for complicated natural structures and processes are being devised and tested. The focal point of our attention is the connection between the chemistry of metal ions and their role for life.

No doubt, we are only at the brink of this process. Thus, it is with the intention of linking coordination chemistry and biochemistry in their widest sense that the series METAL IONS IN BIOLOGICAL SYSTEMS reflects the growing field of "bioinorganic chemistry." We hope, also, that this series will help to break down the barriers between the historically separate spheres of chemistry, biochemistry, biology, medicine, and physics, with the expectation that a good deal of the future outstanding discoveries will be made in the interdisciplinary areas of science.

Should this series prove a stimulus for new activities in this fascinating "field" it would well serve its purpose and would be a satisfactory result for the efforts spent by the authors.

Helmut Sigel

iii

PREFACE TO VOLUME 14

The importance of the so-called trace elements for our health and well-being is now clearly recognized. In regard to metal ions J. M. Wood concluded: "If you think that biochemistry is the organic chemistry of living systems, then you are misled; *biochemistry is the coordination chemistry of living systems*" [*Naturwissenschaften, 62*, 357 (1975)]. It is hoped that the present volume will stimulate further work in this area; it deals with the interplay between metal ions or their complexes and certain manifestations or observations in the field of medicine.

Medicines are often administered orally and then the site of drug absorption is the gastrointestinal tract; therefore the introductory chapter describes drug-metal ion interactions in the gut. Other chapters deal with the deficiency of zinc or iron, as well as with the pharmacological use of these elements. The anti-inflammatory activities of copper and gold complexes are covered, and the role of metal ions and chelating agents in antiviral chemotherapy is considered. Our knowledge of complexes of hallucinogenic drugs is scarce although many of these substances offer binding sites for metal ions; the present contribution has been conceived to promote work in this area. The volume terminates with an evaluation of the use of lithium in psychiatry.

Helmut Sigel

CONTENTS

Chapter 3

THE PHARMACOLOGICAL USE OF ZINC   57

George J. Brewer

Chapter 4

THE ANTI-INFLAMMATORY ACTIVITIES OF COPPER COMPLEXES   77

John R. J. Sorenson

Chapter 5

IRON-CONTAINING DRUGS   125

David A. Brown and M. V. Chidambaram

Chapter 6

GOLD COMPLEXES AS METALLO-DRUGS

Kailash C. Dash and Hubert Schmidbaur

Chapter 7

METAL IONS AND CHELATING AGENTS IN ANTIVIRAL
CHEMOTHERAPY

D. D. Perrin and Hans Stünzi

Chapter 8

COMPLEXES OF HALLUCINOGENIC DRUGS

Wolfram Hänsel

CONTRIBUTORS

Numbers in parentheses indicate the pages on which the authors' contributions begin.

Nicholas J. Birch[*] Department of Biochemistry, The University of Leeds, Leeds, England (257)

George J. Brewer Departments of Human Genetics and Internal Medicine, University of Michigan Medical School, Ann Arbor, Michigan (57)

David A. Brown Department of Chemistry, University College, Dublin, Ireland (125)

M. V. Chidambaram Department of Chemistry, The Florida State University, Tallahassee, Florida (125)

P. F. D'Arcy Department of Pharmacy, The Queen's University of Belfast, Medical Biology Centre, Belfast, Northern Ireland (1)

Kailash C. Dash[†] Institute of Inorganic Chemistry, Technical University of Munich, Garching, Federal Republic of Germany (179)

Wolfram Hänsel Institute for Pharmacy and Food Chemistry, University of Würzburg, Würzburg, Federal Republic of Germany (243)

J. C. McElnay Department of Pharmacy, The Queen's University of Belfast, Medical Biology Centre, Belfast, Northern Ireland (1)

D. D. Perrin Medical Chemistry Group, The John Curtin School of Medical Research, The Australian National University, Canberra, Australia (207)

Ananda S. Prasad Wayne State University School of Medicine and Harper-Grace Hospital, Detroit, Michigan; and Veterans Administration Hospital, Allen Park, Michigan (37)

Present Affiliations:

[*] Department of Biological Sciences, The Polytechnic, Wolverhampton, England

[†] Department of Chemistry, Utkal University, Vani Vihar, Bhubaneswar, India

xi

Hubert Schmidbaur Institute of Inorganic Chemistry, Technical
    University of Munich, Garching, Federal Republic of Germany
    (179)

John R. J. Sorenson Department of Biopharmaceutical Sciences,
    College of Pharmacy and Department of Pharmacology, College of
    Medicine, University of Arkansas for Medical Sciences, Little
    Rock, Arkansas (77)

Hans Stünzi Medical Chemistry Group, The John Curtin School of
    Medical Research, The Australian National University, Canberra,
    Australia (207)

CONTENTS OF OTHER VOLUMES

*Out of print

Other volumes are in preparation.

Comments and suggestions with regard to contents, topics, and the
like for future volumes of the series would be greatly welcome.

METAL IONS IN BIOLOGICAL SYSTEMS

VOLUME 14

Inorganic Drugs in Deficiency and Disease

Chapter 1

DRUG-METAL ION INTERACTIONS IN THE GUT

P. F. D'Arcy and J. C. McElnay
Department of Pharmacy
The Queen's University of Belfast
Medical Biology Centre
Belfast, Northern Ireland

1. INTRODUCTION

Drug interactions can occur at many different sites in the body,
notably in the liver and in the kidneys; drugs may also compete at
receptor or plasma protein binding sites. However, an important
area of drug interaction is at the site of drug absorption in the
gastrointestinal tract. Drug interactions can occur at this site
for a number of reasons, including changes in gastric pH, altered
motility of the gut, interaction with foodstuffs, interaction with
antacid preparations, and the formation of insoluble and nonabsorbable complexes or chelates. Drug-induced malabsorption syndromes
can also be precipitated; for example, Dobbins et al. [1] showed
malabsorption of carotene, hexose sugars, iron, and vitamin B_{12} in
the presence of neomycin, polymyxin, kanamycin, and bacitracin.

In this present chapter consideration is given to drug absorption interactions with metal ions; interactions involving metal-
containing antacids have also been included, although in the latter
malabsorption of drug is often due to changed gastrointestinal pH,
adsorption phenomena, or changed gastric emptying rate rather than
to ionic interactions between metal ions and the drug.

This chapter has been divided into four sections in which
specific aspects of absorption interactions are considered. Firstly,
methods of screening for possible interactions are described, and
this is followed by an account of some interactions that have given
rise to clinical problems. Thirdly, the effects of drug absorption
on pharmacokinetic drug profiles are examined. Finally, discussion
is included on the clinical significance of drug absorption interactions.

2. IN VITRO MODELS OF DRUG ABSORPTION
INTERACTIONS

Drug interactions which give rise to changed drug absorption, especially with self-prescribed medicaments like antacids, are probably a common but rarely recognized cause of therapeutic failure. One approach to learn more about such interactions is by the use of model systems to predict combinations of drugs which will give rise to absorption problems. When such data are known it is then possible to advise the patient to avoid concomitant ingestion of the interacting medicaments or alternatively to advise spacing between their administration by 3 or 4 hr; this should allow normal absorption to take place for each medicament.

Although in vitro models are most useful in the screening of drugs for possible interaction, if an interaction is established by such a procedure, then the clinical significance, if any, of the interaction should be assessed in volunteer subjects or otherwise established in clinical studies.

There are already a range of techniques available for the in vitro study of drug absorption, and many of these can be readily adapted for studying absorption interactions. A brief account of the more useful of these techniques is given below.

2.1. Spectrophotometric Techniques

These methods are useful for the examination of the presence of metal ion-drug complexes. Spectral characteristics of the parent drug can be compared with that of the drug-ion complex. Using this technique the interaction between anhydrotetracycline and various metal ions was established by Stoel et al. [2]. A comparison of the ultraviolet, visible, and fluorescent characteristics of anhydrotetracycline and its metal ion complexes indicated that C_{11} was involved in the metal ion binding. Secondary binding was observed in the A ring by circular dichroism when the primary site was blocked (Fig. 1).

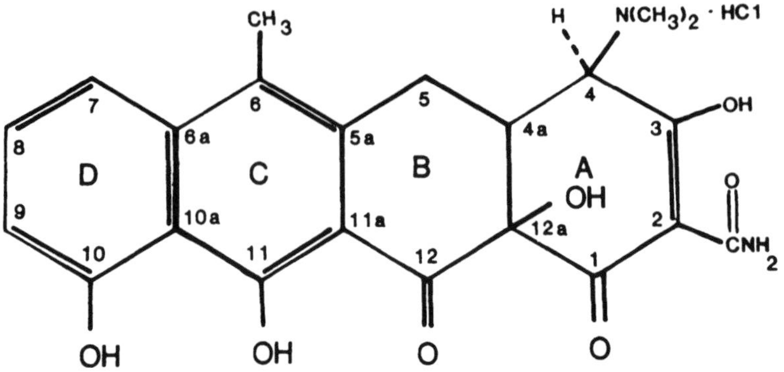

FIG. 1. Structural formula of anhydrotetracycline. (Modified from Ref. 2.)

It is now well known that tetracycline and metal ions interact by chelation; however, this example served to illustrate the potential use of spectral techniques to indicate whether an interaction is involved between two suspected agents. Circular dichroism spectra were also used by Newman and Frank [3] to follow complex formation of tetracycline with Ca^{2+} and Mg^{2+}. Their results indicated that calcium formed a 2:1 metal-ion-to-ligand complex and that the magnesium complex formed at a 1:1 ratio.

A further example of the use of ultraviolet spectrophotometry was made by Chapron et al. [4] during the examination of the phenytoin-calcium interaction. In their study no spectral changes were observed when calcium chloride was added to a phenytoin solution, thus showing that no ionic interaction had taken place. Further discussion is given to the phenytoin-calcium interaction later in this chapter.

2.2. Aqueous/Organic Phase Partitioning

Simple partitioning of drugs between aqueous and organic phases has led to the development of more complicated models; for example, that described by Schulman and Rosano [5]. The underlying principle of models of this type is the separation of two aqueous phases, representing plasma and gastrointestinal lumen, by a liquid oil phase which simulates the lipid nature of the biological membrane [6]. A two-fingered flask, in vitro model which simulated some factors involved in the absorption process was described by Doluisio and Swintosky [7] in 1964. It consisted of a tube containing two aqueous phases separated by an immiscible phase. A rocking apparatus agitated the fluids, causing the liquid interfaces to expand and contract. The results for salicylic acid, barbital, antipyrine, and aminopyrine indicated that initial drug transfer simulated a first-order process. Tetracycline, however, did not undergo transfer from one aqueous phase to the other.

Perrin [8] described a three-phase cell system (aqueous-organic-aqueous) and suggested that his apparatus had advantages in that there was little chance of emulsion formation because of lack of disturbance at the interface and also that samples could be easily removed. In this cell, drug was transferred from a buffer of pH found in the gut, through an organic immiscible liquid, acting as the membrane, to a buffer at the pH of plasma (7.4).

A more recent design of apparatus, is the pear-shaped flask of Koch [9] which utilized a three-phase system. All the systems that have been described have the potential for the study of drug absorption and therefore also for screening drugs for absorption interactions. A limitation of these methods, and indeed one that is shared by all in vitro techniques, is that they cannot allow for changed absorption due to changed gastric emptying rates or altered gastric motility. For example, aluminum and lanthanum are known to have a relaxant effect on gastric smooth muscle [10]; this gives rise to a decreased gastric emptying rate, and this has been implicated in decreased drug absorption [11]. A further disadvantage is that the models cannot account for changes in active drug absorption.

2.3. Rat Everted Intestine Technique

This method is based on the use of everted segments of rat intestine. The use of everted intestinal sacs was first introduced in 1954 by Wilson and Wiseman [12]. Since then the technique has been modified several times by other investigators [13-15]. D'Arcy et al. [16] found the technique useful for the examination of the interaction of tetracyclines with iron and calcium ions; the results obtained in their in vitro study mirrored the clinical data of Neuvonen et al. [17]. The technique described by D'Arcy et al. [16] involved the cannulation of an everted intestinal segment into a tubular system. Two such systems were prepared (control and test) using consecutive segments from the same rat (Fig. 2). The segments were bathed in tissue chambers containing drug solution; the test chamber also contained the examined interactant. Drug absorbed across the intestinal

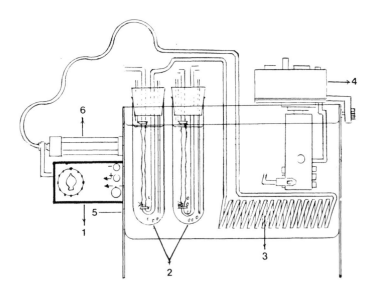

FIG. 2. Schematic diagram of everted rat intestine perfusion apparatus: (1) perfusion pump; (2) intestinal perfusion chambers (boiling tubes); (3) heating coil (glass); (4) thermostated heating unit (37°C); (5) water bath; (6) 50-ml syringe filled with buffer solution.

segments was collected by the infusion of buffer (10 ml) through the tubular system and collection of the sample at the outlet point. The sample contained drug which had been absorbed across the intestine during the previous 10-min period. Control and test data were compared to indicate whether an interaction had occurred.

The everted intestine model has been used successfully to examine the interaction between some drugs and metal-containing antacid constituents, for example tetracyclines [16], digoxin [18], warfarin [19], phenytoin [20], propranolol [21], chloroquine [22], and pyrimethamine [22].

The everted mouse intestine has also been used to examine the effect of cations and other agents on the intestinal transfer of tetracyclines [23].

2.4. Techniques Utilizing Synthetic or Semisynthetic Membranes

Model membranes have been employed in attempts to develop in vitro systems whose transport characteristics correlate with in vivo passive drug absorption [6]. Black (bimolecular) lipid membranes, for example, are primarily used to study transport phenomena and have been used as a model of intestinal absorption of drugs [24]. A commercially available instrument, the Sartorius Absorption Simulator, utilizes synthetic membranes. As well as being useful for drug absorption studies, these systems also lend themselves to the study of drug absorption interactions.

2.5. In Situ Techniques

These methods generally involve the use of anesthetized animals with an intact blood supply, although isolated perfused intestine models are also available for similar studies.

In 1969, Doluisio and co-workers [25] reported a method for the study of gastrointestinal absorption utilizing isolated segments

of the rat small intestine in situ. They reported that their results were closely reproducible and gave absorption rates which were
realistic in terms of known absorption behavior of drugs in humans
and in intact animals.

2.6. Buccal Partitioning

The buccal absorption technique [26] can be considered to be an in
vivo model of drug absorption although it bears many of the hall-
marks of an in vitro method. McElnay and Temple [27] have recently
suggested that this model has a potential for the study of drug absorption interactions involving those drugs which partition into the
oral mucosa. They examined the propranolol-aluminum hydroxide gel
absorption interaction [28] and their results indicated that interaction did not take place due to the formation of a nonabsorbable
complex between the two agents. They therefore suggested that the
observed in vivo interaction was likely to be due to a delayed
gastric emptying rate caused by the aluminum ions of the antacid.

The use of intact laboratory animals can create conditions
which are close to the in vivo situation in humans, and these animal
models have been widely used by a number of investigators [29-31].
It is clear, however, that all nonhuman experimental systems can at
best only be used as screening devices for drug absorption interactions. After interaction between two agents--for example, a metal
ion and a drug--has been indicated in vitro, then the extent of the
interaction must be carefully assessed in patients or in volunteer
subjects. It is only with in vivo studies in humans that the true
extent of bioavailability changes can be found.

Models can, however, be particularly useful if good correlation between in vitro and in vivo data has been firmly established.
In such cases these models can be used with confidence in the laboratory during the screening of medicament combinations whose pharmacology may suggest possible adverse reactions or interactions due to
such combination.

3. INTERACTIONS WITH METAL-CONTAINING ANTACIDS OR METAL SALTS

A diverse range of drugs are involved in drug absorption interactions with antacids. Such interactions can occur by a number of mechanisms--for example, by altering pH (giving rise to changed drug dissolution and absorption patterns); by adsorbing the drugs onto their surface; by altering gastric motility and emptying rate; or via the formation of poorly soluble salts, complexes, or chelates. A wide range of drugs has been implicated in this type of interaction (Table 1), in particular the tetracyclines, phenytoin, digoxin, and chloroquine. In view of the potential importance of such participation, an account of each of the interactions with these drugs and with a miscellaneous group of other drugs is presented in this chapter.

3.1. Tetracyclines

The absorption of tetracyclines and the effect on this of concomitantly administered drugs has been the subject of research by many groups over a considerable period of time. For example, the work of Welch et al. [32] in 1957 demonstrated that a tetracycline-sodium metaphosphate mixture gave higher serum concentrations of tetracycline than did an equivalent amount of tetracycline hydrochloride. This finding encouraged Sweeney et al. [33] to examine the effect of various fillers and adjuvants on the absorption of orally administered tetracycline. One of the most important and interesting findings from that study was that dicalcium phosphate depressed the gut absorption of tetracycline. This finding has since been supported by many other reports; indeed, it is now firmly established that decreased tetracycline absorption results from its interaction not only with Ca^{2+}, but also with many other metal ions, notably Fe^{2+}, Mg^{2+}, and Al^{3+}.

TABLE 1

Drug Absorption Interactions Involving
Metal Salts or Metal-Containing
Antacid Preparations

Drug involved	Interactant	Result of interaction	Reference
Ampicillin	Magnesium hydroxide	Peak serum ampicillin levels depressed more than 50% in mice by concomitant antacid	31
Aspirin	Magnesium and aluminum carbonates	Buffering agents significantly increase rate of aspirin dissolution from solid dosage giving faster peak plasma levels	82
Atenolol	Aluminum hydroxide	Bioavailability of atenolol reduced by average 33%	83
Atropine	Aluminum hydroxide magmas	Adsorption of atropine onto antacid	84
Chlordiaze-poxide	Magnesium/ aluminum hydroxide gel	Decreased rate of absorption of orally administered chlordiazepoxide	85
Chlorproma-zine	Magnesium/ aluminum hydroxide gel	Although it has been reported [86] that the antacid mixture decreased mean urinary concentrations of chlorpromazine, Pinell et al. [87] showed no clear indication of changed serum concentrations; magnesium-trisilicate-aluminum hydroxide mixture has been reported [88] to diminish blood concentrations of chlorpromazine as compared with controls	87
Cimetidine	Aluminum/ magnesium hydroxide	Reduced bioavailability of cimetidine in presence of the antacids; mean decrease 22% in 9 patients	89
Cimetidine	Magnesium trisilicate; magnesium, and aluminum hydroxides	In in vitro studies, Langmuir and Freundlich adsorption isotherms showed cimetidine adsorption was significant with magnesium trisilicate,	90

TABLE 1 (Continued)

Drug involved	Interactant	Result of interaction	Reference
Cimetidine (continued)		but nonexistent with magnesium or aluminum hydroxides; these results contrast with the in vivo findings above [89]	
Dexamethasone	Magnesium trisilicate	Significant decrease in dexamethasone absorption; probably due to its adsorption onto surface of antacid	91
Digoxin	Magnesium perhydrol	Digoxin decomposed by hydrogen peroxide liberated from perhydrol by gastric juice	71
Indomethacin	Aluminum/ magnesium hydroxide	Decreased bioavailability of indomethacin	92
Isoniazid	Aluminum hydroxide gel	Decrease in peak blood levels of isoniazid; probably due to decreased gastric emptying caused by the aluminum hydroxide	77
Levodopa	Magnesium/ aluminum hydroxide	Enhanced levodopa absorption in humans	93
Metoprolol	Aluminum hydroxide	Bioavailability of metoprolol increased by average 11%	83
Nalidixic acid	Sodium bicarbonate	Sodium bicarbonate increased absorption of nalidixic acid and raised its urinary concentration in rabbits	94
Oral contraceptives	Dried aluminum hydroxide gel, magnesium trisilicate	The two antacids adsorb norethisterone acetate; this may affect its absorption and also its contraceptive action	95
Pentobarbitone	Aluminum hydroxide, aluminum salts, and trivalent cation lanthanum	Antacids retarded gastrointestinal absorption of sodium pentobarbitone, lowering its concentration in blood and preventing or delaying sleep in rats	30

TABLE 1 (Continued)

Drug involved	Interactant	Result of interaction	Reference
Phenytoin sodium	Calcium sulfate	Change in capsule formula diluent from calcium sulfate to lactose caused phenytoin intoxication in patients; mechanism of interaction uncertain	54
Propranolol	Aluminum hydroxide gel	Antacid caused decreased bio-availability of propranolol in 4 out of 5 male adults	28
Tetracyclines	Di- and tri-valent metal ions	Ca^{2+}, Mg^{2+}, Fe^{2+}, and Al^{3+} may interact with drugs in the gut, especially tetracyclines, to produce insoluble and nonabsorbable complexes	96
Theophylline	Magnesium/aluminum hydroxide	Antacid significantly decreased absorption rate of theophylline but did not influence extent of absorption or elimination rate	80

Note: Examples are listed in alphabetical order and not in degree of importance.

The accepted reason for the decreased absorption of tetracyclines while in the presence of metal ions is that poorly absorbed chelates are formed. Chelation has been defined as the formation of a compound between a metallic ion and an organic molecule having two groups spatially arranged so as to form a ring structure with the metal [34]. The most stable of tetracycline-metal complexes are, in order of decreasing stability: Fe^{3+}, Al^{3+}, Cu^{2+}, Ni^{2+}, Fe^{2+}, Co^{2+}, Zn^{2+}, and Mn^{2+} [35]. The stoichiometry of the chelation depends on the ion; e.g., calcium forms a 2:1 metal ion:ligand complex, while magnesium forms a 1:1 complex [3]. The diffusion rate of tetracycline is reduced on chelation [36] and this may be one reason why tetracycline chelates are poorly absorbed.

Not only are tetracyclines affected by the presence of metallic ions in coconsumed drugs or in dosage form excipients, they are also malabsorbed when taken with foodstuffs which contain metal ions. Milk and other dairy products (butter, cheese) contain considerable quantities of calcium ions; indeed, the serum concentrations of demeclocycline have been shown to be decreased by some 80% if the antibiotic is taken with milk as compared with the same dosage swallowed with water [37]. Buttermilk decreased the absorption to a lesser degree than did fresh milk; cottage cheese also gave rise to much reduced tetracycline absorption.

Concurrently administered medications containing metal ions, notably antacid mixtures, also seriously affect the absorption of tetracyclines. Aluminum hydroxide is perhaps the best documented interactant in this respect [38-40]; however, many, if not all, antacids and other drugs containing either aluminum, magnesium, or calcium will impair tetracycline absorption, the degree of inhibition of absorption depending on the solubility and stability of the complex formed as well as on the doses used [35]. In this respect it is interesting to note that sodium bicarbonate, an antacid which does not contain polyvalent cations and thus cannot form chelates with tetracyclines, can also significantly reduce tetracycline absorption. The results of studies involving tetracycline-sodium bicarbonate interactions are consistent with a mechanism of absorption requiring low gastric pH for complete dissolution of tetracycline and hence its absorption [41]. Increased pH in the presence of other metal-containing antacids probably also plays a role in the decreased bioavailability of tetracyclines when given together with antacid mixtures.

One of the most striking reductions in tetracycline has implicated iron salts, and this was reported by Neuvonen et al. [17]. In a single-dose study, the serum concentrations of oxytetracycline and tetracycline were reduced by 40-60%, and those of doxycycline and methacycline by 80-90% when the antibiotics were given together with ferrous sulfate (Fig. 3). It must not be forgotten in this

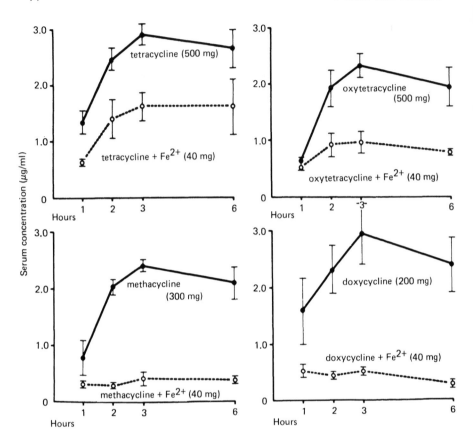

FIG. 3. Effect of concomitant administration of ferrous sulfate (40 mg Fe^{2+}) on the serum concentrations of tetracyclines; both drugs given orally. (From Ref. 17.)

respect that not only is tetracycline absorption much reduced but the absorption of iron is also greatly decreased and the anemia for which the iron is administered will remain untreated or at best considerably undertreated.

The tetracycline-iron interaction is dependent upon the nature of the iron salt used in the combination; Neuvonen and Turakka [42] found that the inhibition of tetracycline absorption by iron was as follows: with ferrous sulfate 85%; with ferrous fumarate, succinate, or gluconate 70-80%; with ferrous tartrate 50%; and with ferric sodium edetate 30%. The dissolution rate of the iron preparation is important in such an interaction; if the iron dissolves slowly, it will liberate ferrous or ferric ions slowly and such delay will allow the tetracycline more time to be effectively absorbed.

The concomitant clinical use of iron salts and tetracyclines by oral dosage need not, however, present problems, since the interaction can be avoided by spacing out the dosage of each component and allowing an interval of at least 3 hr between each medicament [43].

Zinc also has been shown to inhibit the absorption of tetracycline from the gut. The extent of this interaction is shown in Fig. 4, and from the data presented it appears that zinc sulfate is a more potent interactant than the zinc citrate complex [44].

There have been several reports suggesting that the absorption of tetracycline may be changed when in the presence of ethylenediaminetetraacetic acid (EDTA) [45,46]. Recently, Poiger and

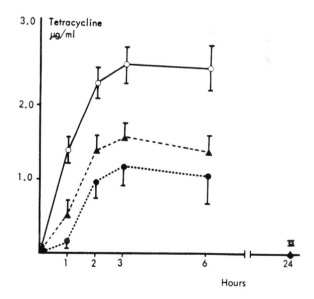

FIG. 4. Serum tetracycyline concentrations (mean ± SEM) in seven healthy volunteers after administration of 500 mg tetracycline (o-o), 500 mg tetracycline + 45 mg Zn^{2+} as a solution of zinc sulfate (●---●), and 500 mg tetracycline + 45 mg Zn^{2+} as a zinc citrate complex (▲---▲). (From Ref. 44.)

Schlatter [47] showed that EDTA did not affect the absorption of tetracycline, but that the decreased absorption of the antibiotic in the presence of milk could be overcome by the simultaneous administration of EDTA. This chelator, it was suggested, prevented complex formation with the tetracycline by itself chelating any available divalent cations. The authors of the report suggested that a tetracycline-EDTA combination might have clinical advantage.

Further work by Poiger and Schlatter [48] has led to the suggestion of an alternative mechanism for absorption interactions with tetracyclines. They suggested that binding of tetracycline to organic macromolecules within the gut is increased by calcium ions and probably also by other di- and trivalent ions; this, they thought, was the main reason for the impaired permeation across the absorbing epithelium. Certainly this latter explanation of the interaction mechanism appears to be more meaningful than one based simply on chelation. It is also supported by the work of Kakemi et al. [46] who showed that the tetracycline-Ca^{2+} complex was more lipophilic than tetracycline alone and hence should be more easily (in contrast to the actual result of the interaction) absorbed.

3.2. Phenytoin

The low solubility of phenytoin in gastrointestinal fluids makes it particularly susceptible to bioavailability problems. With phenytoin, which exhibits Michaelis-Menten kinetics, changes in drug absorption are particularly problematic due to the dose-dependent metabolism and elimination of the drug; this means that small changes in drug absorption may give rise to striking changes in phenytoin plasma concentrations. The situation is also worsened by the fact that phenytoin has a narrow range of therapeutic effectiveness (10-20 µg/ml) [49].

Calcium sulfate significantly decreases the intestinal absorption of phenytoin; this interaction was detected after an outbreak of phenytoin intoxication in a number of Australian cities [50-53].

The cause of the intoxication was subsequently traced to a particular manufacturer who had changed the excipient in 100-mg phenytoin-sodium capsules from calcium sulfate dihydrate to lactose; there had also been a slight increase in the content of both magnesium silicate and magnesium stearate in the capsule formulation [54]. Direct evidence of the interaction was obtained in one patient whose medication was changed from a phenytoin-lactose preparation to a phenytoin-calcium sulfate preparation and finally back to a lactose-containing product. His blood concentration of phenytoin fell sharply when the calcium sulfate-containing product was introduced and rose again when the lactose-containing capsule was reinstated (Fig. 5) [54]. Further studies showed that the calcium sulfate had apparently interacted with the phenytoin to convert about 25% of the anticonvulsant into a form which was insoluble in chloroform and therefore was unlikely to be absorbed from the gut.

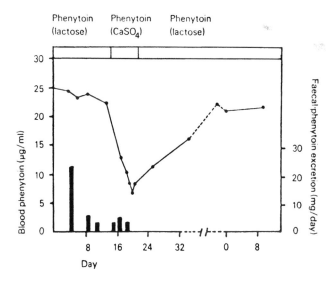

FIG. 5. Blood phenytoin concentrations in a patient taking phenytoin (400 mg/day), with excipients respectively as shown (lactose, calcium sulfate, lactose). Vertical columns represent daily fecal excretion of phenytoin when measured. (From Ref. 54, as presented by Neuvonen, Ref. 60.)

Using a stirred flask method, Bastami and Groves [55] investigated the dissolution behavior of formulations containing phenytoin sodium. When the lactose was replaced by calcium sulfate dihydrate as a diluent in the formulation, there was a reduction in
the release rate of phenytoin although there was no observed change
in disintegration. The release of phenytoin from the calcium sulfate preparation was incomplete, with only 85-90% of the drug going
into solution. Bastami and Groves [55] also examined the effect of
magnesium sulfate on the release of phenytoin and demonstrated complete interference; no phenytoin was released. They suggested that
phenytoin might be forming insoluble salts with calcium and magnesium ions. Earlier work by Glazko and Chang [56] in 1972 had already indicated that phenytoin was capable of forming a chelate with
copper but not with calcium, magnesium, or ferrous ions.

Recent work by Chapron et al. [4] failed to demonstrate an
interaction between calcium or magnesium ions and phenytoin using
ultraviolet absorption spectra while in the presence of calcium
chloride in 0.01N hydrochloric acid. An attempt to produce insoluble complexes of ionized phenytoin with calcium and magnesium ions
at alkaline pH values (0.01N sodium hydroxide) also failed. Their
findings thus implied that complexes with the ionized form of phenytoin either did not form or were extremely soluble in alkaline media.

Antacids have also been implicated in drug absorption interactions with phenytoin; for example, simultaneous ingestion of
antacids has been associated with low plasma phenytoin concentrations in patients, and phenytoin levels were increased two- or
threefold when the same dosage of phenytoin was given 2 or 3 hr
before antacid intake [57].

The fact that antacids (either physician-prescribed or self-
prescribed) are frequently taken by patients prompted Kulshrestha
et al. [58] to study the effects of the concomitant administration
of two antacid preparations, one containing calcium carbonate and
the other a mixture of aluminum hydroxide and magnesium trisilicate,
on phenytoin plasma concentrations in 12 male volunteers. There was

no change in mean serum phenytoin concentrations in the subjects
taking calcium carbonate; however, a slight but significant fall in
the serum concentration of the anticonvulsant was noted in those
who took the aluminum hydroxide-magnesium trisilicate mixture.
The work of O'Brien et al. [59], using aluminum hydroxide gel
and a magnesium hydroxide mixture, and that of Chapron et al. [4],
with a magnesium and aluminum hydroxide or an aluminum hydroxide-
magnesium trisilicate mixture indicated no changes in phenytoin bioavailability when it was given with these antacids. Furthermore,
the latter workers showed that calcium gluconate did not affect
phenytoin bioavailability.

It seems, therefore, that the effects of antacids on the bioavailability of phenytoin are relatively slight. In some patients,
however, this type of interaction is thought to explain the poor
absorption of phenytoin [60].

Herishanu and co-workers [61] suggested that the calcium content of the diet might modify the absorption of phenytoin; they
speculated that phenytoin absorption could be altered by the formation of an insoluble calcium-phenytoin complex or, alternatively, by
the direct effect of calcium ions on the intestinal membrane causing
a "membrane-tightening" effect [62].

With regard to the effect of phenytoin itself on metal ion absorption in the gut, Weismann et al. [63] have shown in rat studies
that the anticonvulsant increases the absorption of zinc. Studies
in epileptic patients by Blyth and Stewart [64] has shown that
phenytoin does not affect serum levels of magnesium.

An overview of the whole area of research suggests that,
although there has been much speculation about the interaction
between phenytoin and metal ions, either in formulation excipients,
antacids, or foodstuffs, firm evidence of interaction in vivo has
only been produced for a phenytoin-calcium sulfate combination.
The mechanism of this interaction still remains uncertain.

3.3. Digoxin

Digoxin, like phenytoin, is also particularly susceptible to bioavailability problems since it has a narrow therapeutic range (0.5
to 2.0 ng/ml) [65].

Binnion [66] has shown that about 20% of digoxin was adsorbed
by the antacid Maalox; the uptake of digoxin and digitoxin by some
antacids has also been reported by Khalil [67]. Magnesium trisilicate showed the greatest adsorptive effect, this being of the order
of 99% for the two glycosides. Other antacids showed a relatively
weaker adsorptive effect; the extent of this did not exceed 25%.

Later work was carried out in volunteer subjects by Brown and
Juhl [68]; cumulative 6-day urinary digoxin excretion data, expressed as a percentage recovery of a 0.75-mg dose, were as follows:
control, 40.1 ± 3.0 (SE); with aluminum hydroxide, 30.7 ± 2.9; with
magnesium hydroxide, 27.1 ± 2.4; and with magnesium trisilicate,
29.1 ± 1.7. The differences in mean values between control and test
data were highly significant, and the authors concluded that the
antacids severely reduced the oral absorption of the digoxin. They
also commented that although the mechanism of this interference with
absorption "remains to be delineated," it could not entirely be related to physical adsorption to the antacids or to alteration in gut
transit time.

The in vitro study of McElnay et al. [18,69], using the
everted rat intestine technique [16], showed that the transfer of
digoxin across the gut was decreased in the presence of aluminum
hydroxide gel (11.4%), by bismuth carbonate (15.2%), by light magnesium carbonate (15.3%), and by magnesium trisilicate (99.5%). In
contrast to these decreased values for digoxin absorption, which
were most likely due to adsorption phenomena, the study by Cooke and
Smith [70] showed that the in vivo absorption of digoxin was unchanged when the glycoside was administered with magnesium trisilicate in the clinical setting. It seems, therefore, that adsorbed
digoxin can be "desorbed" from the antacid during absorption in vivo.

An interaction between digoxin and magnesium perhydrol is interesting because the decreased digoxin bioavailability that results is due to an unusual mechanism. Digoxin is decomposed by hydrogen peroxide which is liberated from magnesium perhydrol on exposure to gastric juice [71].

3.4. Chloroquine

Chloroquine is commonly used in the prophylaxis and treatment of malaria, and its use is still extensive in countries where malaria is endemic; in the Sudan, for example, its use accounts for some 90% of malaria chemotherapy. Self-medication with antacids is common in the Sudan, and also probably in other countries where antimalarials are taken. It is therefore very likely that these two classes of medication will be taken in combination. The results of a recent study by McElnay et al. [22], using the everted rat intestine model, have shown that the intestinal transfer of chloroquine is changed when in the presence of specific antacids (Table 2); changes in absorption patterns for another antimalarial, pyrimethamine, were also demonstrated in the presence of antacids (Table 2) [22].

These studies have been extended, and presently good correlation between in vitro data (rat intestine model) and in vivo data has been obtained in six healthy Negro-Arab volunteers who took chloroquine alone and then together with magnesium trisilicate. The other antacids listed in Table 2 have not yet been tested in vivo for their effect on chloroquine absorption.

Findings of interactions between antimalarials and other co-consumed drugs is especially worrying since with many such agents the prophylactic dosage against malaria is taken only once per week. If absorption of this dosage is reduced, then prophylactic cover against malaria may be lost until the next dose is taken 1 week later.

TABLE 2

Effects of Various Antacid Constituents
on the Transfer of Chloroquine and
Pyrimethamine across Everted
Rat Intestine

Constituent	Change in chloroquine transfer, %	Change in pyrimethamine transfer, %
Magnesium trisilicate	-31.3 S	-37.5 S
Calcium carbonate	-52.8 S	-31.5 VHS
Sodium bicarbonate	+68.7 S	+ 0.9 NS
Gerdiga[a]	-36.1 S	-38.0 VHS
Aluminum hydroxide gel	+31.0 HS	+31.9 VHS

Key: VHS = P < 0.001; HS = P < 0.01; S = P < 0.05; NS = P > 0.05.

[a]Gerdiga is a hydrated silicate similar to atapulgite and is used as
an antacid in outlying areas of the Sudan.

3.5. Miscellaneous Drugs

3.5.1. Oral Anticoagulants

It has been suggested, purely on theoretical grounds, that the absorption of warfarin from the gut may be changed by antacids [72].
Clinical studies, however, have not confirmed this prediction;
Robinson et al. [73] showed that an antacid mixture of aluminum and
magnesium hydroxides did not alter plasma warfarin concentrations.
Ambre and Fisher [74] examined the effect of the same antacids on
the absorption of anticoagulants; they showed that peak plasma
levels of bishydroxycoumarin were increased by 75% when it was given
with magnesium hydroxide. Bishydroxycoumarin concentrations were
not affected by the coadministration of aluminum hydroxide; warfarin absorption was not affected by either of the antacids.

These latter workers also showed that a magnesium chelate of bishydroxycoumarin was absorbed more rapidly than its sodium salt; this suggested that the increased absorption of this anticoagulant in the presence of magnesium hydroxide was due to chelate formation. Later work by Bighley and Spivey [75] showed that this chelate has a 2:1 ligand-metal stoichiometry with 2 moles of water associated with the complex.

Small decreases in the absorption of warfarin in the in vitro everted intestine model have been noted by McElnay et al. [19] when the anticoagulant was tested in the presence of bismuth carbonate (6.9%) and magnesium trisilicate (19.3%). This latter result with magnesium trisilicate suggests that this interaction might present instability problems in warfarin-treated patients if concomitant therapy is used.

3.5.2. Nitrofurantoin

Naggar and Khalil [76] have demonstrated that magnesium trisilicate reduced the rate and the extent of nitrofurantoin excretion which reflected a decrease in the rate and extent of its absorption. The time during which nitrofurantoin concentration in urine was above the minimum effective level of 32 μg/ml was also significantly reduced; adsorption of nitrofurantoin onto the surface of the antacid appeared to be the most probable mechanism of this interaction. Magnesium trisilicate adsorbed 99.3% of the nitrofurantoin in that study, while bismuth oxycarbonate, aluminum hydroxide, magnesium oxide, and calcium carbonate gave rise, respectively, to adsorption values of 53.3, 2.5, 26.8, and 0%.

3.5.3. Sulfonamides

Animal (rat) and human studies have indicated that concomitant treatment with antacids, which raise gastric pH sufficiently to increase dissolution of the acid form of the sulfonamide, hastens the absorption of sulfonamides since their absorption rate is limited by the amount of drug in solution [77].

The information presented in this present section on the effects of antacids on the absorption and kinetics of drugs has, of necessity, been abbreviated. More detailed information may be gained from two excellent and comprehensive reviews published by Romankiewicz [78] in 1976 and Hurwitz [77] in 1977.

4. IMPLICATIONS OF ABSORPTION INTERACTIONS ON DRUG PHARMACOKINETIC PROFILES

Drug absorption interactions can alter the bioavailability of a given dose of drug; such interaction can affect both the rate as well as the extent of drug absorption.

4.1. Rate of Drug Absorption

Assuming a first-order absorption process, the rate of drug absorption after oral dosage will depend on the rate at which the drug is presented at the site of absorption, normally the small intestine. This can be influenced by the rate of drug release from the dosage form (e.g., disintegration), the dissolution of the drug in the gut contents, and also the rate of stomach emptying. In drug interactions involving metal ions or metal-containing antacids, the rate of release and dissolution of a drug may be influenced by pH, as for example when antacids are used, while the rate of gastric emptying may be delayed by aluminum ions which cause gastrointestinal relaxation [79].

4.2. Extent of Drug Absorption

The extent of drug absorption, which is measured as AUC_{∞}^{0} (area under the curve), depends on the amount of drug released from its dosage form and presented at the site of absorption in an absorbable form. The classical interaction between tetracycline and di- or trivalent

metal ions is an example of this; the amount of tetracycline available for absorption is decreased due to complex formation involving
metal ions.

Changes in the rate and/or extent of drug absorption can give
rise to changes in the pharmacokinetic parameters t_{max} (time taken
to reach maximum or peak plasma level) and Cp_{max} (peak plasma level)
after a single dose of drug. The effect of changed absorption patterns on these two parameters have been considered in Table 3. It
is clear from these data that nine combinations of extent and rate
of drug absorption are possible, although certain types of interaction are more probable than others. The most important interactions
are those which give rise to increased or decreased drug absorption
since they may result in corresponding increased or decreased total
pharmacological effects in the multiple-dose situation. Changed
rate of drug absorption can give rise to changed Cp_{max} values; increased rates will evoke larger than normal fluctuations in plasma
drug concentrations during multiple-dose therapy although generally
changes will only be of clinical significance if the drug involved
has a narrow range of therapeutic effectiveness. In such circumstances the Cp_{max} may reach toxic levels.

Increased extent of drug absorption will only be possible for
drugs which have low intrinsic bioavailability, or when the drug--
for example, propranolol--is susceptible to extensive first-pass
metabolism. In this latter event, an increased rate of absorption
may be sufficient to saturate this first-pass effect and allow more
drug to become available for systemic circulation.

Drug absorption interactions should not, however, be considered
in isolation; the pharmacodynamics (e.g., therapeutic index) of the
primary drug involved and changes in the pharmacokinetic parameters
outlined (e.g., intrinsic bioavailability, absorption rate, and
first-pass metabolism) must also be taken into account. It is only
then that it can be predicted, with any degree of certainty, whether
an absorption interaction is likely to evoke clinically significant
problems.

TABLE 3

Effects of Changed Absorption Patterns on
Drug Plasma Profiles and Drug Interactions

Type	AUC_∞^0 (extent of drug absorption)	Rate of drug absorption	Resulting t_{max}	Resulting Cp_{max}	Possibility of interaction
1.	Increased	Increased	Decreased	Increased	Yes
2.	Decreased	Decreased	Increased	Decreased	Yes
3.	No change	No change	No change	No change	No
4.	Increased	Decreased	Increased	Uncertain	Yes
5.	Decreased	No change	No change	Decreased	Yes
6.	No change	Increased	Decreased	Increased	Yes
7.	Increased	No change	No change	Increased	Yes
8.	Decreased	Increased	Decreased	Uncertain	Yes
9.	No change	Decreased	Increased	Decreased	Yes

Some examples of interactions which have given rise to changed drug pharmacokinetic plasma profiles are as follows. Firstly, the ingestion of dairy products influences the absorption of orally administered tetracyclines; Scheiner and Altemeier [37] showed that the lowered serum concentration of the antibiotic that resulted from such interaction could result in failure of treatment (Fig. 6). Secondly, Arnold et al. [80] showed that plasma theophylline concentrations "without antacid" and those "with antacid" differed significantly, but only at 0.67 and 1 hr after dosage (Fig. 7). Although the absorption rate of theophylline was significantly lower in the presence of the antacid than without it, no significant differences were found in the AUC data. These authors therefore concluded that concurrent administration of the magnesium-aluminum hydroxide gel antacid (Maalox) with Aminophyllin brand theophylline ethylenediamine tablets would not be expected to alter the clinical efficacy of the theophylline.

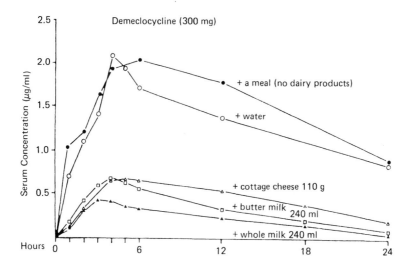

FIG. 6. Effect of some dairy products or a meal without dairy products on the absorption of demeclocycline (demethylchlortetracycline). Mean serum concentrations in four to six subjects. (Data from Ref. 37, by permission of *Surgery, Gynecology and Obstetrics* as presented by Neuvonen, Ref. 35.)

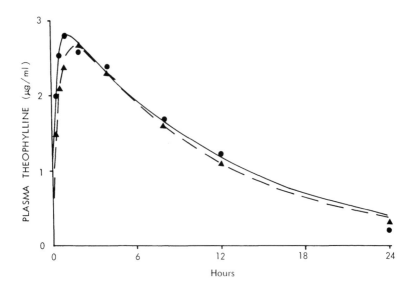

FIG. 7. Mean plasma theophylline levels of patients receiving theophylline alone (●) and theophylline plus antacid (▲) fitted to
theoretical one-compartment bioavailability curves (theophylline
alone ———, theophylline with antacid ---). (From Ref. 80.)

A third example has already been mentioned (Sec. 3.5.2);
Naggar and Khalil [76] assessed absorption characteristics by considering the cumulative amount of drug excreted in the urine and
showed (Fig. 8) that magnesium trisilicate decreased both the rate
and the extent of nitrofurantoin excretion. This indicated both a
decreased rate and a decreased extent of drug absorption.

The solution to some absorption interaction problems can be
quite simple, as has been demonstrated by Gothoni et al. [43] who
showed that interaction between oral iron and tetracyclines can be
avoided by simply spacing the separate administration of these drugs
by 2 or 3 hr (Fig. 9). If the spacing was shorter than this, then
problems arose; for example, iron taken 1 hr after tetracycline
caused a 40% reduction in the absorption of the antibiotic. If
taken 2 or 3 hr *after* the antibiotic, then plasma concentrations of
tetracycline were within expected levels. Interestingly, iron taken
up to and including 2 hr *before* the tetracycline significantly reduced the plasma concentration of the antibiotic.

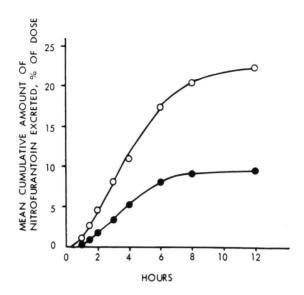

FIG. 8. Mean cumulative amount of nitrofurantoin excreted in the urine as a function of time after oral administration of nitrofurantoin tablet (100 mg) alone and when administered with 5 g of magnesium trisilicate (lower curve). (From Ref. 76, V. F. Naggar and S. A. Khalil, *Clin. Pharmacol. Ther.*, *25*, 857, 1979.)

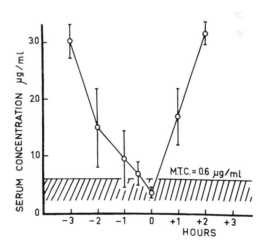

FIG. 9. Graphic presentation of the iron-tetracycline interaction. Ordinate: mean serum concentration 3 hr after a single dose of 500 mg tetracycline. Abscissa: interval between intake of tetracycline and 600 mg ferrous sulfate (left side: iron taken before; right side: iron taken after tetracycline). (From Ref. 43.)

5. CLINICAL SIGNIFICANCE OF DRUG ABSORPTION
    INTERACTIONS WITH METAL IONS AND
    METAL-CONTAINING ANTACIDS

In general, drug interactions of this type have been overstated and
overestimated; in vitro studies and results from animal experiments
have frequently and wrongly been extrapolated to the clinical scene.
A simple example of this is that in an in vitro experiment a specific
drug may appear to be adsorbed by an antacid agent. This may also
occur when the combination is administered to humans; however, it is
of little importance if the adsorbed drug is rapidly and effectively
"desorbed" from the antacid during its passage down the gut. The
result of the treatment is thus not impaired and the primary drug
is, in practice, just as effective when given with antacid as when
given alone.

     Models of absorbent or adsorbent interactions are useful, but
clinical studies or trials in volunteer subjects can provide the only
definitive judgment of whether the interaction is really of any
importance clinically.

     A second generalization is that absorption interactions are
only likely to create clinical emergency when the primary drug interactant has a low therapeutic index. In such a case, small alterations in
drug absorption may evoke large changes in therapeutic effect. The
drug may reach a toxic level or it may not provide effective treatment; these are the outside limits of such interaction.

     It is important to distinguish clearly between rate of absorp-
tion and extent of absorption. It is likely, for example, that the
decreased gastric emptying rate caused by aluminum ions could cause
a decreased absorption rate of any drug taken in combination with an
aluminum-containing antacid. The interaction would generally only
become important if the extent of the absorption of a drug was
adversely affected.

     Differences in total absorption may be of little clinical sig-
nificance if the drug is administered well in excess of requirements;
this is so for many of the antibiotics. In such circumstances only

very large decreases in absorption are likely to change the ultimate therapeutic effect of the dosage [81].

Even if drug absorption interactions are significant and are of potential clinical importance, it is salutory to realize that they can often be avoided without discontinuing one or other of the interactants. Often all that is required is that the two interacting drugs are given at a spaced interval so that they do not meet "headlong" in the gut; mention has already been made of the work of Gothoni et al. [43] who established that iron salts taken 2 or 3 hr after tetracyclines did not result in impaired tetracycline absorption.

A final generalization is that physicians should always consider a drug absorption interaction as a reason for unexpected therapeutic failure; nonprescribed medication, especially antacids, may be interacting with prescribed drugs. The patient should always be warned not to take any other drugs, including household remedies, or other self-prescribed medicines when they are taking prescribed drugs unless they check with the doctor or pharmacist first that such combinations of treatment are not likely to interact. It is important, in this respect, not only to educate physicians and pharmacists on the potential hazards of drug interactions, but also to instruct the patients as well.

REFERENCES

1. W. O. Dobbins, B. A. Herrero, and C. M. Mansbach, *Amer. J. Med. Sci.*, *225*, 63 (1968).

2. L. J. Stoel, E. C. Newman, G. L. Asleson, and C. W. Frank, *J. Pharm. Sci.*, *65*, 1794 (1976).

3. E. C. Newman and C. W. Frank, *J. Pharm. Sci.*, *65*, 1728 (1976).

4. D. J. Chapron, P. A. Kramer, S. L. Mariano, and D. C. Hohnadel, *Arch. Neurol.*, *36*, 436 (1979).

5. J. H. Schulman and M. Rosano, in *Physics of Interfaces*, Proceedings of the 3rd International Congress of Surface Activity, Cologne, September 1960. Vol. 2, Universitäts-Druckerei, Mainz, 1961. p. 112.

6. K. A. Herzog and J. Swarbrick, *J. Pharm. Sci.*, *59*, 1759 (1970).

7. J. T. Doluisio and J. V. Swintosky, *J. Pharm. Sci.*, *53*, 597 (1964).

8. J. Perrin, *J. Pharm. Pharmacol.*, *19*, 25 (1967).

9. H. Koch, *Dtch. Apoth. Zeit.*, *117*, 1241 (1977).

10. M. Hava and A. Hurwitz, *Eur. J. Pharmacol.*, *22*, 156 (1973).

11. A. Hurwitz and D. L. Schlozman, *Amer. Rev. Resp. Dis.*, *109*, 41 (1974).

12. T. H. Wilson and G. Wiseman, *J. Physiol.*, *123*, 116 (1954).

13. E. G. Lovering and D. B. Black, *J. Pharm. Sci.*, *63*, 671 (1974).

14. Z. T. Chowan and A. A. Amaro, *J. Pharm. Sci.*, *66*, 1249 (1977).

15. J. Blanchard and A. Straussner, *Acta Pharm. Suec.*, *14*, 279 (1977).

16. P. F. D'Arcy, H. A. Muhyiddin, and J. C. McElnay, *J. Pharm. Pharmacol.*, *Suppl.*, *28*, 33P (1976).

17. P. J. Neuvonen, G. Gothoni, R. Hackman, and K. Björksten, *Brit. Med. J.*, *4*, 532 (1970).

18. J. C. McElnay, D. W. G. Harron, P. F. D'Arcy, and M. R. G. Eagle, *Experientia*, *35*, 94 (1979).

19. J. C. McElnay, D. W. G. Harron, P. F. D'Arcy, and P. S. Collier, *Experientia*, *35*, 1359 (1979).

20. J. C. McElnay, *Brit. J. Pharmacol.*, *70*, 126P (1980).

21. J. C. McElnay, P. F. D'Arcy, and J. K. Leonard, *Irish J. Med. Sci.*, *149*, 448 (1980).

22. J. C. McElnay, H. A. Mukhtar, P. F. D'Arcy, and D. J. Temple, unpublished observations (1980).

23. S. Banerjee and K. Chakrabarti, *J. Pharm. Pharmacol.*, *28*, 133 (1976).

24. K.-I. Inui, K. Tabara, R. Hori, A. Kaneda, S. Muranishi, and H. Sezaki, *J. Pharm. Pharmacol.*, *29*, 22 (1977).

25. J. T. Doluisio, N. F. Billups, L. W. Dittert, E. T. Sugita, and J. V. Swintosky, *J. Pharm. Sci.*, *58*, 1196 (1969).

26. A. H. Beckett and E. J. Triggs, *J. Pharm. Pharmacol.*, *Suppl.*, *19*, 31S (1967).

27. J. C. McElnay and D. J. Temple, *Drug Intell. Clin. Pharm.*, *15*, 481 (1981).

28. J. H. Dobbs, V. A. Skoutakis, S. R. Acchiardo, and B. R. Dobbs, *Curr. Ther. Res.*, *21*, 887 (1977).

29. W. R. Adam and J. K. Dawborn, *Aust. N.Z. J. Med.*, *2*, 126 (1971).

30. A. Hurwitz and M. B. Sheehan, *J. Pharmacol. Exp. Ther.*, *179*, 124 (1971).

31. A. Hurwitz, C. Rauscher, and S. H. Wan, *Clin. Res.*, *21*, 818
     (1973).

32. H. Welch, W. W. Wright, and A. Kirshbaum, *Antibiot. Med.*, *4*,
     293 (1957).

33. W. M. Sweeney, S. M. Hardy, A. C. Dornbush, and J. M. Ruegseg-
     ger, *Antibiot. Med.*, *4*, 642 (1959).

34. S. P. Bessman and N. J. Doorenbos, *Ann. Intern. Med.*, *47*, 1036
     (1957).

35. P. J. Neuvonen, *Drugs*, *11*, 45 (1976).

36. T.-F. Chin and J. L. Lach, *Amer. J. Hosp. Pharm.*, *32*, 625
     (1975).

37. J. Scheiner and W. A. Altemeier, *Surg. Gynec. Obstet.*, *114*, 9
     (1962).

38. B. A. Waisbren and J. S. Hueckel, *Proc. Soc. Exp. Biol. Med.*,
     73, 73 (1950).

39. J. E. Rosenblatt, J. E. Barrett, J. L. Brodie, and W. M. M.
     Kirby, *Antimicrob. Agents Chemother.*, *6*, 134 (1966).

40. G. Levy, in *Prescription Pharmacy*, 2nd ed. (J. B. Sprowls, Jr.,
     ed.), Lippincott, Philadelphia, 1970, pp. 70, 75, 80.

41. W. H. Barr, J. Adir, and L. Garrettson, *Clin. Pharmacol. Ther.*,
     12, 779 (1971).

42. P. J. Neuvonen and H. Turakka, *Eur. J. Clin. Pharmacol.*, *7*,
     357 (1974).

43. G. Gothoni, P. J. Neuvonen, M. Mattila, and R. Hackman, *Acta
     Med. Scand.*, *191*, 409 (1972).

44. K.-E. Andersson, L. Bratt, C. Dencker, C. Kamme, and E. Lanner,
     Eur. J. Clin. Pharmacol., *10*, 59 (1976).

45. G. M. Eisenberg, W. Weiss, and H. F. Flippin, *J. Lab. Clin.
     Med.*, *52*, 895 (1958).

46. K. Kakemi, H. Sezaki, H. Ogata, and T. Nadai, *Chem. Pharm.
     Bull.* (Tokyo), *16*, 2206 (1968).

47. H. Poiger and C. Schlatter, *Eur. J. Clin. Pharmacol.*, *14*, 129
     (1978).

48. H. Poiger and C. Schlatter, *Nauyn-Schmiedeberg's Arch. Phar-
     makol. Exp. Pathol.*, *306*, 89 (1979).

49. H. Kutt and F. McDowell, *J. Amer. Med. Ass.*, *203*, 167 (1968).

50. L. Rail, *Med. J. Aust.*, *2*, 339 (1968).

51. J. Balla, *Med. J. Aust.*, *2*, 480 (1968).

52. M. J. Eadie, J. M. Sutherland, and J. H. Tyrer, *Med. J. Aust.*,
     2, 515 (1968).

53. P. J. Landy, *Med. J. Aust.*, *2*, 639 (1968).

54. J. H. Tyrer, M. J. Eadie, J. M. Sutherland, and W. D. Hooper, *Brit. Med. J.*, *4*, 271 (1970).

55. S. M. Bastami and M. J. Groves, *Int. J. Pharmaceut.*, *1*, 151 (1978).

56. A. J. Glazko and T. Chang, in *Antiepileptic Drugs* (D. M. Woodbury, J. K. Penry, and R. P. Schmidt, eds.), Raven Press, New York, 1972, pp. 127 ff.

57. H. Kutt, *Epilepsia*, *16*, 393 (1975).

58. V. K. Kulshrestha, M. Thomas, J. Wadsworth, and A. Richens, *Brit. J. Clin. Pharmacol.*, *6*, 177 (1978).

59. L. S. O'Brien, M. L'E. Orme, and A. M. Breckenridge, *Brit. J. Clin. Pharmacol.*, *6*, 176 (1978).

60. P. J. Neuvonen, *Clin. Pharmacokinet.*, *4*, 91 (1979).

61. J. Herishanu, V. Eylath, and R. Ilan, *Israel J. Med. Sci.*, *12*, 1453 (1976).

62. J. F. Manery, in *Mineral Metabolism* (C. L. Comar and F. Bronner, eds.), Academic Press, New York, 1969, pp. 441 ff.

63. K. Weismann, L. Knudsen, and H. Høyer, *J. Invest. Derm.*, *71*, 396 (1978).

64. A. Blyth and J. Stewart, *Clin. Chim. Acta*, *62*, 305 (1975).

65. V. P. Butler and J. Lindenbaum, *Amer. J. Med.*, *58*, 460 (1975).

66. P. F. Binnion, in *Symposium on Digitalis* (O. Storstein Glylendal, ed.), Norsk. Förlag, Oslo, 1973, pp. 216 ff.

67. S. A. H. Khalil, *J. Pharm. Pharmacol.*, *26*, 961 (1974).

68. D. D. Brown and R. P. Juhl, *New Engl. J. Med.*, *295*, 1034 (1976).

69. J. C. McElnay, D. W. G. Harron, P. F. D'Arcy, and M. R. G. Eagle, *Brit. Med. J.*, *1*, 1554 (1978).

70. J. Cooke and J. A. Smith, *Brit. Med. J.*, *2*, 1166 (1978).

71. W. J. F. van der Vijgh, J. H. Fast, and J. E. Lunde, *Drug Intell. Clin. Pharm.*, *10*, 680 (1976).

72. D. A. Hussar, *J. Amer. Pharm. Ass.*, *10*, 78 (1970).

73. D. S. Robinson, D. M. Benjamin, and J. J. McCormack, *Clin. Pharmacol. Ther.*, *12*, 491 (1971).

74. J. J. Ambre and L. J. Fisher, *Clin. Pharmacol. Ther.*, *14*, 231 (1973).

75. L. D. Bighley and R. J. Spivey, *J. Pharm. Sci.*, *66*, 1124 (1977).

76. V. F. Naggar and S. A. Khalil, *Clin. Pharmacol. Ther.*, *25*, 857 (1979).

77. A. Hurwitz, *Clin. Pharmacokinet.*, *2*, 269 (1977).

78. J. A. Romankiewicz, *Primary Care*, *3*, 537 (1976).

79. A. Hurwitz, G. G. Robinson, T. S. Vats, F. C. Whittier, and W. F. Herrin, *Gastroenterology*, *71*, 268 (1976).

80. L. A. Arnold, G. H. Spurbeck, W. H. Shelver, and W. M. Henderson, *Amer. J. Hosp. Pharm.*, *36*, 1059 (1979).

81. G. G. Graham and M. Kennedy, *Med. J. Aust.*, *2*, 143 (1978).

82. R. K. Nayak, R. D. Smyth, A. Polk, T. Herczeg, V. Carter, A. J. Visalli, and N. H. Reavey-Cantwell, *J. Pharmacokinet. Biopharm.*, *5*, 597 (1977).

83. P. Lundborg, in *Proceedings of the International Conference on Drug Absorption*, Edinburgh, 1979, p. 105.

84. I. W. Grote and M. Woods, *J. Amer. Pharm. Ass. (Sci. Ed.)*, *42*, 319 (1953).

85. D. J. Greenblatt, R. I. Shader, J. S. Harmatz, K. Franke, and J. Koch-Weser, *Clin. Pharmacol. Ther.*, *19*, 234 (1976).

86. F. M. Forrest, I. S. Forrest, and M. T. Serra, *Biol. Psychiat.*, *2*, 53 (1970).

87. O. C. Pinell, D. C. Fenimore, C. M. Davis, O. Moreira, and W. E. Fann, *Clin. Pharmacol. Ther.*, *23*, 125 (1978).

88. W. E. Fann, J. M. Davis, D. S. Janowsky, H. J. Sekerke, and D. M. Schmidt, *J. Clin. Pharmacol.*, *13*, 388 (1973).

89. G. Bodemar, B. Norlander, and A. Walan, *Lancet*, *i*, 444 (1979).

90. F. Ganjian, A. J. Cutie, and T. Jochsberger, *J. Pharm. Sci.*, *69*, 352 (1980).

91. V. F. Naggar, S. A. Khalil, and W. Gouda, *J. Pharm. Sci.*, *67*, 1029 (1978).

92. R. L. Galeazzi, *Eur. J. Clin. Pharmacol.*, *12*, 65 (1977).

93. L. Rivera-Calimlim, C. A. Dujovne, J. O. Morgan, L. Lasagna, and J. R. Bianchine, *Eur. J. Clin. Invest.*, *1*, 313 (1971).

94. E. J. Segre, H. Sevelius, and J. Varaday, *New Eng. J. Med.*, *291*, 582 (1974).

95. H. Fadel, A. Abdelbary, E. Nour El-Din, and A. A. Kassem, *Pharmazie*, *34*, 49 (1979).

96. J. P. Griffin and P. F. D'Arcy, *A Manual of Adverse Drug Interactions*, 2nd ed., Wright, Bristol, 1979, pp. 3 ff.

Chapter 2

ZINC DEFICIENCY AND ITS THERAPY

Ananda S. Prasad
Wayne State University School of Medicine
and Harper-Grace Hospital
Detroit, Michigan
and
Veterans Administration Hospital
Allen Park, Michigan

1. HISTORY

Nearly 100 years ago, zinc was shown to be essential for microorganisms. In 1934 two groups of workers showed convincingly that zinc was required for the growth and well-being of rats [1]. A disease called parakeratosis of swine was related to zinc deficiency in 1955 [2], and it was shown to be essential for the birds in 1958. Clinical manifestations in zinc-deficient animals include growth retardation, testicular atrophy, skin changes, and poor appetite.

In the fall of 1958, a 21-year-old patient at Saadi Hospital, Shiraz, Iran who looked like a 10-year-old boy was brought to my attention. In addition to dwarfism, he had severe anemia, hypogonadism, hepatosplenomegaly, rough and dry skin, mental lethargy, and geophagia [3]. He ate only bread made of wheat flour, and the intake of animal protein was negligible. He also consumed nearly 1 lb of clay daily. The habit of geophagia (clay eating) is common in the villages around Shiraz. Ten additional similar cases came to our attention in a short period of time.

The anemia was due to iron deficiency even though no source of blood loss could be documented. We concluded that this may have been attributable to a lack of availability of iron from the bread, greater iron loss due to excessive sweating in hot climate, and adverse effect of geophagia on iron absorption. In every case, the anemia was corrected by oral iron therapy.

It was difficult to explain all the clinical features solely on the basis of tissue iron deficiency, inasmuch as growth retardation and testicular atrophy are not seen in iron-deficient experimental animals. The possibility that zinc deficiency may be present was considered. Zinc deficiency in animals was shown to produce growth retardation, testicular atrophy, and skin changes. Inasmuch as some metals may form insoluble complexes with phosphates, we speculated that factors responsible for decreased availability of iron in these patients may also affect the availability of zinc adversely.

Similar patients were encountered in Egypt subsequently [4,5]. Their dietary history was also similar, except that geophagia was not documented. These patients were documented to have zinc deficiency. This conclusion was based on the following: (1) the zinc concentrations in plasma, red cells, and hair were decreased; and (2) radioactive zinc-65 studies revealed that the plasma zinc turnover rate was greater, the 24-hr exchangeable pool was smaller, and the excretion of zinc-65 in stool and urine was less in the patients than in the control subjects.

Further studies in Egypt showed that the rate of growth was greater in patients who received supplemental zinc compared with those who received only animal protein diet consisting of bread, beans, lamb, chicken, eggs, and vegetables [6-8]. Pubic hair appeared in all cases within 7-12 weeks after zinc supplementation was initiated. Genitalia size became normal, and secondary sexual characteristics developed within 12-24 weeks in all subjects receiving zinc. On the other hand, no such changes were observed in a comparable length of time in the iron-supplemented group or in the group on an animal protein diet alone. Thus, the growth retardation and gonadal hypofunction in these subjects was related to zinc deficiency.

Later studies in Iran demonstrated that zinc is a principal limiting factor in the nutrition of children [9]. It was also evident that the requirement of zinc under different dietary conditions varied widely in the Middle East. For instance, in the studies reported by Prasad [6] and Sandstead et al. [8], 18 mg of supplemental zinc with adequate animal protein and calories intake was sufficient to produce a definite response with respect to growth and gonads, but in other studies when the subjects continued to eat the village diet, up to 40 mg of zinc supplement was required to show some growth effect [9].

In another study from Iran, Halsted et al. [10] showed that the development in 19- or 20-year old subjects receiving a well-balanced diet alone was slow, and the effect on height increment and onset of sexual function was strikingly enhanced in those receiving zinc. The two women described in their report were from a hospital clinic and represented the first cases of dwarfism in females due to zinc deficiency.

2.   CAUSES OF HUMAN ZINC DEFICIENCY

2.1.   Nutritional

Nutritional deficiency of zinc in humans is fairly prevalent throughout the world.   Besides Iran and Egypt, it has now been reported from Turkey, Portugal, Morocco, Yugoslavia, and other developing countries [11].   Zinc deficiency should be prevalent in other countries where primarily cereal proteins are consumed by the population.

Plasma zinc levels have been measured in infants and young children suffering from kwashiorkor in Cairo, Pretoria, Capetown, and Hyderabad [12].   At the time of hospital admission, plasma zinc levels were very low in all four locations, but the hypozincemia could be attributed at least in part to hypoalbuminemia.   At a time of "clinical cure" when total serum protein and albumin levels were normal, in Cairo and Pretoria, plasma zinc levels remained significantly low, suggesting that zinc deficiency is associated with kwashiorkor in some geographical locations.   After recovery from kwashiorkor, children in some areas of the world, including Egypt, are likely to receive a diet inadequate in zinc, and a deficiency of this nutrient may persist indefinitely, thus contributing to growth failure.

Nutritional deficiency of zinc occurring in children and infants has been recognized also in the United States.   A study from Denver in 1972 identified a number of children from middle- and upper-income families who had a significantly decreased zinc level

in the hair and retarded growth [13]. A later study showed that
hair and plasma zinc levels were exceptionally low in infants in the
United States [14]. Several factors, including difficulty in achiev-
ing positive zinc balance in early postnatal life and a "dilutional"
effect of rapid growth, may contribute to zinc depletion in infants.
A unique factor in the United States which may contribute to zinc deficiency is the low concentration of this element in certain popular
infant milk formulas [15]. Zinc supplementation of all of these
formulas has not been routine, though it is likely to become a uni-
versal practice in the near future following the recent recommenda-
tions of the Food and Nutrition Board of the National Academy of
Sciences with respect to zinc intake in infants. It appears that
those infants at the lower end of a spectrum of zinc depletion, as
manifested by low levels of hair and plasma zinc, and perhaps those
who remain moderately depleted for a prolonged period of time, do
develop symptomatic zinc deficiency [15].

2.2. Alcohol

Alcohol induces hyperzincuria [16,17]. The mechanism is not well
understood. A direct effect of alcohol on renal tubules may be
responsible for hyperzincuria.

The serum zinc level of the alcoholic subjects tends to be
lower in comparison with the controls. An absolute increase in
renal clearance of zinc in the alcoholics demonstrable at both normal level and low serum zinc concentration has been observed [18].
Thus, the measurement of renal clearance of zinc may be clinically
used for etiological classification of chronic liver disease attributable to alcohol in different cases.

Excessive ingestion of alcohol may lead to a severe deficiency
of zinc, as reported by Weismann et al. [19]. In this case, widespread eczema craquele, hair loss, steatorrhea, dysproteinemia with
edema, and mental disturbances due to zinc deficiency were observed
[19]. Therapy with zinc reversed these manifestations. A similar

clinical syndrome has been seen among Ugandian blacks addicted to banana gin.

2.3. Liver Disease

Vallee et al. [20] demonstrated that patients with cirrhosis of the liver had low serum and hepatic zinc and, paradoxically, hyperzincuria. These observations suggested that zinc deficiency in the alcoholic cirrhotic patient may be a conditioned deficiency which was somehow related to alcohol ingestion. These observations have been now confirmed by several investigators.

Recently, Morrison et al. [21] reported abnormal dark adaptation in six stable alcoholic cirrhotics who also had low serum zinc levels. Zinc supplementation to these patients resulted in improvement of dark adaptation.

2.4. Gastrointestinal Disorders

Zinc deficiency has been reported in patients with steatorrhea [12]. In an alkaline environment, zinc would be expected to form insoluble complexes with fat and phosphates. Thus, fat malabsorption due to any cause should result in an increased loss of zinc in the stool.

Exudation of large amounts of zinc protein complexes into the intestinal lumen also may contribute to the decrease in plasma zinc concentrations which occur in patients with inflammatory disease of the bowel. It seems likely that protein-losing enteropathy may impair zinc homeostatis. Another potential cause of negative zinc balance is a massive loss of intestinal secretions.

2.5. Neoplastic Diseases

The occurrence of conditioned deficiency of zinc in patients with neoplastic diseases will obviously depend upon the nature of the neoplasm. Anorexia and starvation, plus abundance of foods lacking available zinc, are probably important conditioning factors.

2.6. Burns and Skin Disorders

The causes of zinc deficiency in patients with burns include losses in exudates. Starvation of patients with burns is a well-recognized cause of morbidity and mortality. The contribution of conditioned zinc deficiency to the morbidity of burned patients is not defined. Limited studies indicate that epithelialization of burns can be improved by treatment with zinc. Such a finding is consistent with the beneficial effect of zinc on the treatment of leg ulcers and the well-defined requirement of zinc for collagen synthesis [22,23].

2.7. Impaired Wound Healing in Chronically Diseased Subjects

In 1966 Pories and Strain [24] reported that oral administration of zinc to military personnel with marsupialized pilonidal sinuses was attended by a twofold increase in the rate of reepithelialization. The authors' conclusion that zinc promotes wound healing remained controversial for several years. Studies in experimental animals have demonstrated that (1) healing of incised wounds is impaired in rats with zinc deficiency; (2) collagen and noncollagen proteins are reduced in skin and connective tissues of rats with dietary zinc deficiency; and (3) zinc supplementation does not augment wound healing in normal rats. Most reports indicate that zinc supplementation is beneficial for wound healing in zinc-deficient patients.

2.8. Renal Disease

In a recent study, Mahajan et al. [25] have documented that patients with chronic renal failure have low concentration of zinc in plasma,

leukocytes, and hair as well as increased plasma ammonia level and
increased activity of plasma ribonuclease. These biochemical changes
in chronic renal failure suggest that zinc deficiency is a complicat-
ing feature of uremia. It was further concluded that chronic dialysis therapy did not correct zinc deficiency in such subjects. In
later studies by Mahajan et al. [26] it was also observed that the
taste abnormalities and hypogonadism in patients with chronic renal
failure were reversed with zinc supplementation. These studies suggest that zinc deficiency occurs in patients with chronic renal
failure and that zinc supplementation should be considered in order
to correct the deficiency.

2.9. Pregnancy

An additional positive zinc balance of approximately 375 mg is required during normal pregnancy. In one study, mild deficiency of
zinc in pregnant women was reported to be associated with increased
maternal morbidity, abnormal taste sensation, prolonged gestation,
inefficient labor, atonic bleeding, and increased risks to the fetus
[27]. Zinc supplementation to the deficient mother resulted in reduced frequency of the above complications [27]. Undoubtedly, further research and controlled clinical trials with zinc supplementation are indicated in pregnancy in order to confirm the above
observations.

2.10. Iatrogenic Causes

Possible iatrogenic causes of conditioned deficiency of zinc include
use of chelating agents, antimetabolites, antianabolic drugs and
diuretics. Failure to include zinc in fluids for total parenteral
nutrition (TPN) is another example of iatrogenically induced conditioned deficiency of zinc. In some cases the deficiency may be very
severe and resemble congenital type of acrodermatitis enteropathica.

Zinc deficiency occurring in patients following penicillamine therapy for Wilson's disease has been reported recently [28]. The manifestations consisted of parakeratosis, "dead" hair and alopecia, keratitis, and centrocecal scotoma. The clinical features were again very similar to those observed in congenital type of acrodermatitis enteropathica. Zinc supplementation resulted in a complete clinical cure.

2.11. Genetic Disorders

Acrodermatitis enteropathica is a lethal, autosomal, recessive trait which usually occurs in infants of Italian, Armenian, or Iranian lineage. The disease is not present at birth, but typically develops in the early months of life, soon after weaning from breast-feeding. Dermatological manifestations include progressive bullous-pustular dermatitis of the extremities and of the oral, anal, and genital areas, combined with paronychia and generalized alopecia. Infection with *Candida albicans* is a frequent complication. Ophthalmic signs may include blepharitis, conjunctivitis, photophobia, and corneal opacities. Gastrointestinal disturbances are usually severe, including chronic diarrhea, malabsorption, steatorrhea, and lactose intolerance. Neuropsychiatric signs include irritability, emotional disorders, tremor, and occasional cerebellar ataxia. These patients generally have retarded growth and hypogonadism. Zinc supplementation results in complete cure [29,30]. The underlying mechanism of the zinc deficiency in these patients is due to malabsorption. The genetic basis of zinc malabsorption remains to be elucidated.

2.12. Sickle Cell Disease

Recently, deficiency of zinc in sickle cell disease has been recognized [30]. Certain clinical features are common to some sickle cell anemia patients and zinc-deficient patients, the latter as reported from the Middle East [4]. These symptoms include delayed onset of puberty and hypogonadism in the males, short stature, low body weight, rough skin, and poor appetite.

In a study reported by Prasad et al. [30,31], zinc in plasma, erythrocytes, leukocytes, and hair was decreased and urinary zinc excretion was increased in sickle cell anemia patients as compared with controls. Erythrocyte zinc and daily urinary zinc excretion were inversely correlated in the anemia patients ($r = 0.71$, $p < 0.05$), suggesting that hyperzincuria may have caused zinc deficiency in these patients. Carbonic anhydrase, a zinc metalloenzyme, correlated significantly with erythrocyte zinc ($r = 0.94$, $p < 0.001$). Neutrophil alkaline phosphatase (a zinc metalloenzyme) activity is considerably decreased in sickle cell anemia patients and along with neutrophil zinc levels seems to correlate with zinc status [32]. Plasma ribonuclease (RNase) activity was significantly greater in anemia subjects than in controls and inasmuch as zinc is an inhibitor of this enzyme, the observed activity is consistent with the hypothesis that sickle cell anemia patients were zinc-deficient [30,31]. The activity of another zinc-dependent enzyme, deoxythymidine kinase, was observed to be decreased in newly synthesized collagen connective tissue in sickle cell anemia subjects [33].

Zinc supplementation under controlled conditions showed that sickle cell anemia patients gained weight, their serum testosterone level increased, and plasma ammonia level decreased [32]. We also observed abnormal dark adaptation in some sickle cell anemia patients which improved following zinc supplementation [34]. Zinc supplementation also decreased the number of irreversible sickled cells in venous blood, presumably by its action on the red cell membrane; however, at present there is no conclusive evidence that zinc supplementation will prevent pain crisis. Further work is needed in this area.

2.13. Experimental Production of Zinc Deficiency in Humans

Although the role of zinc in human subjects has been now defined and its deficiency recognized in several clinical conditions, these examples are not representative of a pure zinc-deficient state in humans. It was therefore considered desirable to develop a human model which would allow a study of the effects of a mild zinc-deficient state in humans and also would provide sensitive parameters which could be used clinically for diagnosing zinc deficiency. Recently such a model has been established successfully in human volunteers, and changes in several zinc-dependent parameters have been documented [35].

A decrease in zinc concentration of plasma, erythrocytes, leukocytes, and urine and changes in the activities of zinc-dependent enzymes such as plasma alkaline phosphatase, RNase in the plasma, and deoxythymidine kinase in the newly synthesized collagen connective tissue were observed specifically in relation to the dietary intake of zinc [35]. One unexpected finding was with respect to plasma ammonia level, which increased due to dietary zinc restriction but was reversible with zinc supplementation [35]. We have now confirmed this finding also in zinc-deficient rats [36].

Our data indicate that following zinc-deficient state, the subjects showed a greater positive balance for zinc. This would suggest that perhaps a test based on oral challenge of zinc and subsequent plasma zinc determination may be able to distinguish between zinc-deficient and zinc-sufficient states in human subjects.

Dietary zinc restriction resulted in a decrease in body weight in spite of adequate caloric and protein intake [35]. Testicular functions were also affected adversely as a result of restricted zinc intake, but all these effects were reversible with zinc supplementation [37].

3. CLINICAL MANIFESTATIONS OF ZINC DEFICIENCY

Growth retardation, hypogonadism in the males, poor appetite, mental
lethargy, and skin changes were the classical clinical features of
chronically zinc-deficient subjects from the Middle East as reported
by the author in the early 1960s [7]. All the above-mentioned
features were corrected by zinc supplementation. Liver and spleen
were also found to be enlarged in the zinc-deficient dwarfs which
improved following zinc supplementation. The mechanism of spleen
and liver enlargment in this syndrome is not well understood at
present.

Recently abnormal dark adaptation in alcoholic cirrhotics has
been related to a deficiency of zinc [21]. Zinc administration to
these patients resulted in improvement of dark adaptation [21].
Similar clinical observations have now been made in some zinc-
deficient sickle cell anemia patients [34]. It has been proposed
that the effect of zinc on the retina may be mediated by an enzyme
retinene reductase, which is known to be zinc-dependent. This interesting clinical observation needs further documentation.

It is likely that some of the clinical features of cirrhosis
of the liver such as loss of body hair, testicular hypofunction,
poor appetite, mental lethargy, difficulty in wound healing, and
night blindness may indeed be related to the secondary zinc-deficient
state in this disease. In the future, careful clinical trials with
zinc supplementation need to be carried out in order to establish
the effects of zinc in patients with chronic liver disease.

"Fetal alcohol syndrome" is characterized by prenatal and
postnatal growth deficiency, microcephaly, short palpebral fissures,
epicanthal folds, cleft palate, small jaw, joint deformities, and
cardiac, renal, and genitalia anomalies. Many of these features are
similar to those reported in rat fetuses when zinc intake was restricted in the mothers during the crucial stage of gestation [38].
In view of the fact that excessive alcohol intake may deplete body
zinc stores and inasmuch as zinc plays a vital role in DNA synthesis

and cell division, one may speculate that a maternal zinc deficiency
may indeed be responsible for the "fetal alcohol syndrome." A careful clinical study is needed to test this hypothesis.

In sickle cell disease, delayed onset of puberty and hypogonadism in the males, characterized by decreased facial, pubic, and
axillary hair, short stature, and low body weight, rough skin, and
poor appetite have been noted [30-32] and related to a secondary
zinc-deficient state. Many patients with sickle cell disease
develop chronic leg ulcers which do not heal, and a beneficial effect of zinc supplementation in such cases has been reported. Further controlled clinical trials are indicated in order to establish
the effect of zinc therapy in this condition.

Some patients with coeliac disease who failed to respond to
diet, steroids, and nutritional supplements, made remarkable recovery
when zinc was administered. They gained weight, and d-xylose absorption test and steatorrhea improved following zinc therapy [12]. Zinc
supplementation in a few subjects with malabsorption syndrome (other
than coeliac disease) seemed to have produced beneficial results with
respect to growth retardation, hypogonadism in the males, mental
lethargy, skin changes, and loss of hair. One should therefore be
aware of the occurrence of zinc deficiency as a possible complication
of malabsorption syndrome, since this is easily correctable.

The conclusion that zinc can promote the healing of cutaneous
sores and wounds has been the subject of controversy for several
years; however, most studies now provide evidence that zinc supplementation does promote wound healing in zinc-deficient patients and
that zinc therapy in zinc-sufficient subjects is not effective for
wound healing.

Abnormalities of taste have been related to a deficiency of
zinc in humans by some investigators [39-41]. Decreased taste acuity
(hypogeusia) has been observed in zinc-deficient subjects, such as
patients with liver disease, malabsorption syndrome, following
thermal burns, chronic uremia, and following administration of
penicillamine or histidine [39,40]. A double-blind study, however,

failed to show the effectiveness of zinc in the treatment of hypogeusia in various diseases [41]. Recently, however, Mahajan et al.
[26] have reported that zinc was effective in improving taste acuity
in chronic uremia subjects. Their studies were carried out in a
double-blind fashion. This may suggest that depletion of zinc may
lead to decreased taste acuity but that not all cases of hypogeusia
are due to zinc deficiency. The role of zinc in hypogeusia needs to
be delineated further.

The dermatological manifestations of severe zinc deficiency
include progressive bullous-pustular dermatitis of the extremities
and the oral, anal, and genital areas, combined with paronychia and
generalized alopecia such as seen in acrodermatitis enteropathica.
Infection with Candida albicans is a frequent complication.

Neuropsychiatric signs include irritability, emotional disorders, tremors, and occasional cerebellar ataxia. The patients
generally have retarded growth and males exhibit hypogonadism. Zinc
therapy has been shown to produce remarkable improvements and is
considered to be a life-saving measure in these subjects.

A very similar clinical picture has been reported in a patient
who received penicillamine therapy for Wilson's disease. Following
total parenteral nutrition and excessive ingestion of alcohol,
clinical manifestation of zinc deficiency resemble acrodermatitis
enteropathica, and these conditions therefore may be considered
examples of acquired acrodermatitis enteropathica. Once a deficiency
of zinc is recognized, zinc therapy becomes imperative in such cases.

4. LABORATORY DIAGNOSIS OF ZINC DEFICIENCY

Measurement of zinc level in plasma is very useful provided the
sample is not hemolyzed and contaminated. In condition of acute
stress, following myocardial infarction, and acute infections, zinc
from the plasma compartment may redistribute to other tissues, thus
making an assessment of zinc status in the body a difficult task.
Intravascular hemolysis would also increase the plasma zinc level

inasmuch as the content of red cell zinc is much higher than the plasma.

Zinc in the red cell and hair also may be used for assessment of body zinc status; however, inasmuch as these are slowly-turning-over tissues, the zinc levels in these tissues do not reflect recent changes with respect to body zinc stores. Neutrophil zinc determination, on the other hand, appears to reflect the body zinc status more accurately and is thus a very useful parameter [32,35]. Quantitative assay of alkaline phosphatase activity in the neutrophils is also a very useful tool in our experience [32].

Urinary excretion of zinc is decreased as a result of zinc deficiency. Thus, determination of zinc in 24-hr urine may be of additional help in diagnosing zinc deficiency provided cirrhosis of the liver, sickle cell disease, and chronic renal disease are ruled out. These conditions are known to have hyperzincuria and associated zinc deficiency.

In our experiments, during the zinc-deficient state our subjects showed a marked positive balance for zinc [35]. Thus, a metabolic balance study may clearly distinguish zinc-deficient from zinc-sufficient subjects. One may also suggest that perhaps a test based on oral challenge of zinc and subsequent plasma zinc determination may be able to distinguish between zinc-deficient and zinc-sufficient states in human subjects.

The activities of many zinc-dependent enzymes have been shown to be affected adversely in zinc-deficient tissues. Three enzymes, alkaline phosphatase, carboxypeptidase, and thymidine kinase, appear to be most sensitive to zinc restriction in that their activities are affected adversely within 3 to 6 days of institution of a zinc-deficient diet to experimental animals. In human studies, the activity of deoxythymidine kinase in proliferating skin collagen and alkaline phosphatase activity in neutrophils were shown to be sensitive to dietary zinc intake. As a practical test, quantitative measurement of alkaline phosphatase activity in neutrophils may be very useful adjunct to neutrophil zinc level determination in order to assess body zinc status in humans.

5. TREATMENT OF HUMAN ZINC DEFICIENCY

A daily zinc intake of 15 mg for adults has been recommended by the
Food and Nutrition Board of the National Academy of Sciences of the
United States, with an additional 5 mg during pregnancy and 10 mg
during lactation. The daily zinc requirement of preadolescent children has been estimated to be 10 mg. Zinc allowance for infants up
to the age of 6 months has been tentatively set at 3 mg/day.

Nutritional deficiency of zinc in adolescents and adults in
the Middle East was corrected with 18 mg of oral zinc supplementation when the diet was adequate in animal protein. In Iranian villages, however, 40 mg of zinc supplementation only partially corrected zinc deficiency. This was because the villagers continued to
eat only bread with no animal protein intake and inasmuch as cereal
proteins contain high amounts of phytate, an organic phosphate compound, the availability of zinc for absorption was not adequate.
Thus, one must take into consideration the nature of the dietary protein intake in order to assess optimal zinc supplementation for cor-
rection of deficiency.

Several investigators have used therapeutically zinc sulfate
220 mg three times a day for wound healing or 150 mg of zinc in six
divided doses at four hourly intervals for treatment of pain crisis
in sickle cell disease [24,42]. In such amounts, copper deficiency
may result [43]. The major manifestations were microcytosis and
neutropenia, both of which were corrected with copper supplementation
(2 mg/day). A recent report suggests that high-density lipoprotein
cholesterol may decrease with 440 mg of zinc sulfate administration
daily for 5 weeks [43]. It is likely that this effect of zinc was
due to copper deficiency induced by high dosage of zinc administration. It is therefore recommended that zinc supplementation in
amounts larger than recommended dietary allowance should be monitored
carefully.

Acute toxic symptoms such as abdominal pain, nausea, vomiting,
fever, drowsiness, and lethargy have been observed to occur if zinc

is ingested in amounts exceeding much more than 1 g in single doses. Some subjects may experience epigastric discomfort with zinc sulfate capsules; however, this can be eliminated by using zinc acetate or zinc glutamate.

ACKNOWLEDGMENTS

This work was supported in part by a grant from the United States Department of Agriculture, Competitive Research Grants, and a Sickle Cell Center Grant from National Heart and Lung Institute, National Institutes of Health, Bethesda, Maryland.

REFERENCES

1. W. R. Todd, C. A. Elvehjem, and E. B. Hart, *Amer. J. Physiol.*, *107*, 146 (1934).

2. H. F. Tucker and W. D. Salmon, *Proc. Soc. Exp. Biol. Med.*, *88*, 613 (1955).

3. A. S. Prasad, J. A. Halsted, and M. Nadimi, *Amer. J. Med.*, *31*, 532 (1961).

4. A. S. Prasad, A. Miale, Jr., Z. Farid, H. H. Sandstead, and W. J. Darby, *Arch. Int. Med.*, *111*, 407 (1963).

5. A. S. Prasad, A. Miale, Z. Farid, A. Schulert, and H. H. Sandstead, *J. Lab. Clin. Med.*, *61*, 531 (1963).

6. A. S. Prasad, Metabolism of zinc and its deficiency in human subjects, in *Zinc Metabolism* (A. S. Prasad, ed.), Charles C. Thomas, Springfield, Ill., 1966, p. 250.

7. A. S. Prasad, Deficiency of zinc in man and its toxicity, in *Trace Elements in Human Health and Disease* (A. S. Prasad, ed.), Academic Press, New York, 1976, p. 1.

8. H. H. Sandstead, A. S. Prasad, A. R. Schulert, Z. Farid, A. Miale, Jr., S. Bassily, and W. J. Darby, *Amer. J. Clin. Nutr.*, *20*, 422 (1967).

9. H. A. Ronaghy, J. G. Reinhold, M. Mahloudji, P. Ghavami, M. R. S. Fox, and J. A. Halsted, *Amer. J. Clin. Nutr.*, *27*, 112 (1974).

10. J. A. Halsted, H. A. Ronaghy, P. Abadi, M. Haghshenass, G. H. Amirhakimi, R. M. Barakat, and J. G. Reinhold, *Amer. J. Med.*, *53*, 277 (1972).

11. J. A. Halsted, J. C. Smith, Jr., and M. I. Irwin, *J. Nutr.*, *105*, 345 (1974).

12. H. H. Sandstead, K. P. Vo-Khactu, and N. Solomon, Conditioned zinc deficiencies, in *Trace Elements in Human Health and Disease*, Vol. 1 (A. S. Prasad, ed.), Academic Press, New York, 1976, p. 33.

13. K. M. Hambidge, C. Hambidge, M. Jacobs, and J. D. Baum, *Pediatr. Res.*, *6*, 868 (1972).

14. K. M. Hambidge and P. A. Walravens, Zinc deficiency in infants and preadolescent children, in *Trace Elements in Human Health and Disease*, Vol. 1 (A. S. Prasad, ed.), Academic Press, New York, 1976, p. 21.

15. P. A. Walravens and K. M. Hambidge, Nutritional zinc deficiency in infants and children, in *Zinc Metabolism: Current Aspects in Health and Disease* (G. J. Brewer and A. S. Prasad, eds.), Alan R. Liss, New York, 1977, p. 61.

16. J. F. Sullivan, *Quart. J. Stud. Alcohol, 23*, 216 (1962).

17. S. Gudbjarnason and A. S. Prasad, Cardiac metabolism in experimental alcoholism, in *Biochemical and Clinical Aspects of Alcohol Metabolism* (V. M. Sardesai, ed.), Charles C. Thomas, Springfield, Ill., 1969, p. 266.

18. J. G. Allan, G. S. Fell, and R. I. Russel, *Scott Med. J.*, 109 (1975).

19. K. Weismann, J. Roed-Peterson, N. Hjorth, and H. Kopp, *Int. J. Dermatol.*, *15*, 757 (1976).

20. B. L. Vallee, W. E. C. Wacker, A. F. Bartholomay, and E. D. Robin, *N. Engl. J. Med.*, *255*, 403 (1956).

21. S. A. Morrison, R. M. Russell, E. A. Carney, and E. V. Oaks, *Amer. J. Clin. Nutr.*, *31*, 276 (1978).

22. S. L. Husain, *Lancet, 1*, 1069 (1969).

23. F. Fernandez-Madrid, A. S. Prasad, and D. Oberleas, *J. Lab. Clin. Med.*, *82*, 951 (1973).

24. W. J. Pories and W. H. Strain, Zinc and wound healing, in *Zinc Metabolism* (A. S. Prasad, ed.), Charles C. Thomas, Springfield, Ill., 1966, p. 378.

25. S. K. Mahajan, A. S. Prasad, P. Rabbani, W. A. Briggs, and F. D. McDonald, *J. Lab. Clin. Med.*, *94*, 693 (1979).

26. S. K. Mahajan, A. S. Prasad, J. Lambujon, A. Abbasi, W. A. Briggs, and F. D. McDonald, *Amer. J. Clin. Nutr.*, *33*, 1517 (1980).

27. S. Jameson, Zinc and pregnancy, in *Zinc in the Environment*, Part 2, *Health Effects* (J. O. Nriagu, ed.), Wiley, New York, 1980, p. 183.

28. W. G. Klingberg, A. S. Prasad, and D. Oberleas, Zinc deficiency
     following penicillamine therapy, in *Trace Elements in Human
     Health and Disease,* Vol. 1 (A. S. Prasad, ed.), Academic Press,
     New York, 1976, p. 51.

29. E. J. Moynahan, *Proc. R. Soc. Med., 59,* 7 (1966).

30. E. J. Moynahan and P. M. Barnes, *Lancet, 1,* 676 (1973).

31. A. S. Prasad, J. Ortega, G. J. Brewer, and D. Oberleas, *J.
     Amer. Med. Ass., 235,* 2396 (1976).

32. A. S. Prasad, A. Abbasi, P. Rabbani, and E. DuMouchelle, *Amer.
     J. Hematol.,* in press.

33. A. S. Prasad, F. Fernandez-Madrid, and J. F. Ryan, *Amer. J.
     Physiol., 236*(3), E272 (1979).

34. A. S. Prasad, Zinc deficiency and effects of zinc supplementa-
     tion on sickle cell anemia subjects, in *Erythrocyte Structure
     and Function* (G. J. Brewer, ed.), Alan R. Liss, New York, 1981.

35. A. S. Prasad, P. Rabbani, A. Abbasi, E. Bowersox, and M. R. S.
     Fox, *Ann. Intern. Med., 89,* 482 (1978).

36. P. Rabbani and A. S. Prasad, *Amer. J. Physiol., 235,* E203
     (1978).

37. A. Abbasi, A. S. Prasad, P. Rabbani, and E. DuMouchelle, *J.
     Lab. Clin. Med., 96,* 544 (1980).

38. L. S. Hurley, Perinatal effects of trace element deficiencies,
     in *Trace Elements in Human Health and Disease,* Vol. 2 (A. S.
     Prasad, ed.), Academic Press, New York, 1976, p. 301.

39. R. I. Henkin, Zinc, saliva, and taste, in *Zinc and Copper in
     Clinical Medicine* (K. M. Hambidge and B. L. Nichols, eds.),
     Spectrum, Jamaica, N.Y., 1978, p. 35.

40. R. I. Henkin, P. J. Schechter, R. C. Hoye, and C. F. T.
     Mattern, *J. Amer. Med. Ass., 217,* 434 (1971).

41. R. I. Henkin, P. J. Schechter, W. T. Friedewald, D. L. Demets,
     and M. S. Raff, *Amer. J. Med. Sci., 272,* 285 (1976).

42. A. S. Prasad, G. J. Brewer, E. B. Schoomaker, and P. Rabbani,
     J. Amer. Med. Ass., 240, 2166 (1978).

43. P. L. Hooper, L. Visconti, P. J. Garry, and G. E. Johnson, *J.
     Amer. Med. Ass., 244,* 1960 (1980).

Chapter 3

THE PHARMACOLOGICAL USE OF ZINC

George J. Brewer
Departments of Human Genetics and Internal Medicine
University of Michigan Medical School
Ann Arbor, Michigan

1. HISTORY

First we should make it clear that when we speak of zinc as a pharmacological agent, we are discussing its use as a drug beyond simple
replacement of zinc in zinc-deficiency syndromes of the type discussed in the preceding chapter. The pharmacological use of zinc

does not have a long and extensive history. The major early area of
use has been to attempt improvement of wound healing. In addition,
there have been a variety of other minor uses and trials. Our discussion of the history of the use of zinc as a drug will not be ex-
haustive, but will briefly review the more prominent areas.

The major pharmacological use of zinc for a long time was to
improve the healing of wounds. The recorded use of topical zinc to
promote healing dates back to about 1500 B.C. [1]. In 1953, Strain
et al. [2] demonstrated that increase in dietary zinc improved wound
healing in rats. It seems likely that the standard diet in use at
that time provided inadequate zinc, and that the rats were slightly
zinc-deficient [1]. Today's rat chow is normally zinc-enriched, and
additional supplementation beyond this level does not improve wound
healing. A large number of studies have been done on a variety of
animals (reviewed in Refs. 1 and 3) all of which generally bear out
the observation that if the animal is zinc-deficient, supplemental
zinc helps wound healing, whereas supranormal levels do not provide
additional benefit.

In 1967 Pories et al. [4,5] reported beneficial effects of
oral zinc on wound healing in humans. Initial patients studied had
pilonidal cysts. Healing rate was almost doubled in the treated
group versus the control group. A number of studies (reviewed in
Ref. 1) have reported improved healing resulting from oral zinc
therapy in venous stasis leg ulcers, in major burns, in chronic
wounds of a variety of types, and in leg ulcers in sickle cell
anemia [6]. Some of these studies have been carefully controlled.
Some investigators have failed to find an effect of zinc on wound
healing (reviewed in Refs. 1 and 3), and it is now clear that the
usefulness of zinc in this area depends upon the zinc status of the
patient. If the patient is zinc-deficient, zinc supplementation is
helpful, whereas if the patient is zinc-sufficient, additional zinc
is not helpful. Since so many patients with chronic healing problems also are zinc-deficient or at least have a marginal zinc
status, many such patients do benefit from zinc therapy.

The more or less standard dose of zinc administered for wound healing purposes has been 220 mg of zinc sulfate given three times a day with meals. Since each 220 mg of zinc sulfate contains 50 mg of elemental zinc, this daily dose is 150 mg of elemental zinc. While this is a pharmacological dose of zinc, as we have pointed out above, elevation of zinc above normal physiological levels does not further enhance wound healing. Thus, this therapeutic use of zinc, rather than being a pharmacological use, is another example of the therapeutic effect of zinc in zinc deficiency. This use has, however, provided considerable pharmacological experience with zinc.

We will not discuss here the use of pharmacological doses of zinc in diseases such as acrodermatitis enteropathica, cirrhosis of the liver, total parental nutrition, taste and smell deficiencies, etc., since these appear to be additional examples of zinc deficiency syndromes.

In the early 1970s our group proposed the use of pharmacological doses of zinc for its antisickling properties in sickle cell anemia. This should not be confused with the use of zinc in correcting zinc deficiency in sickle cell anemia, which was discussed in the preceding chapter. The proposal to use zinc as an antisickling agent was at first empiric, based upon observations that zinc at fairly low concentrations would enhance the ability of sickled cells to go through small pores in Nucleopore filters [7]. By 1977, it had been shown that oral zinc therapy did indeed have an in vivo antisickling effect, and would lower the count of irreversibly sickled cells (ISCs) in sickle cell patients [8]. The mechanistic hypothesis proposed [7,9] was antagonism by zinc of detrimental effects of increased intracellular calcium [10] in sickle cells. This hypothesis has subsequently been extended to include inhibition of calmodulin function by zinc [11,12]. Calmodulin is a calcium-activated protein, and the concept is that because of excessive calcium in the sickle cell, there is excessive calmodulin activation. The hypothesis for zinc action is that zinc, by virtue of its being an inhibitor of calmodulin function, has a beneficial effect.

One of the by-products of the work in sickle cell anemia has been the development of an alternate regimen for the oral administration of zinc. It has been shown that zinc absorption is significantly inhibited by most of the foods present in a normal meal [13, 14]. Thus zinc given with food is relatively ineffective as a pharmacological agent. Further, the elevated plasma level of zinc, because it lasts for only about 5 hr after a dose, does not permit effective continuous elevation of plasma zinc with three-times-a-day administration. To accomplish this, zinc must be given five or six times a day. The pharmacological use of zinc in sickle cell anemia will be further reviewed in Sec. 4.

In 1976, Simkin [15] showed that oral zinc therapy was of some benefit in rheumatoid arthritis. In a 12-week trial, he showed that oral zinc therapy given by the traditional regimen of 220 mg of zinc sulfate three times a day with meals had a significant inhibitory effect on some of the inflammatory components of the disease such as joint swelling, tenderness, and morning stiffness. To our knowledge, a follow-up and confirmation of these observations in human patients has not been done. We will discuss animal studies in a later section.

2. ZINC ABSORPTION

It should not be too surprising that certain foods inhibit absorption of pharmacologically administered zinc, since the original basis of zinc deficiency in humans was the chelation of zinc by phytates in the unleavened bread diet in certain villages in the Middle East [16]. Thus, it has been understood for some time that phytates and vegetable fibers will chelate zinc and prevent its absorption. However, the extent of this effect in terms of the number of foods which inhibit zinc absorption and the extent to which this inhibition interferes with pharmacological administration of zinc have not been sufficiently appreciated [13,14].

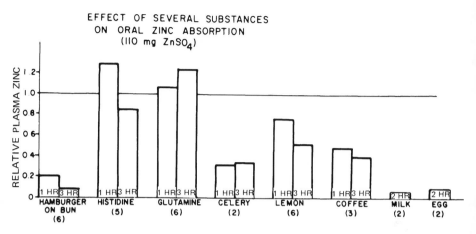

FIG. 1. The effect of foods and certain amino acids on the absorption of orally administered zinc. The line at 1.0 indicates the expected mean plasma zinc response at various times based on studies of oral zinc administration to the same subjects while fasting. The bars indicate the relative plasma zinc response when the same dose of zinc is taken with the indicated substances. For example, when zinc is taken with a hamburger, the mean plasma zinc level at 1 hr is only 0.2 as high in the same subjects as when the same dose of zinc is taken fasting. The number in parentheses indicates the number of subjects. (From Ref. 13.)

The dramatic effect of food on the absorption of pharmacologically administered zinc is illustrated in Fig. 1 [13]. Figure 1 shows the plasma zinc response to a dose of 25 mg of elemental zinc when the zinc is taken with a variety of foods compared to the plasma zinc response when the zinc is taken fasting (line at 1.0). The distance below the 1.0 line of the tops of the columns indicates the degree of inhibition of zinc absorption at 1 and 3 hr after ingestion. Figure 1 illustrates how widespread the phenomenon of food blockage of zinc absorption is and includes foods and drinks such as the bread of a hamburger bun, celery, coffee, tea, milk, egg, and lemon juice, all of which significantly block zinc absorption. Similar studies have been carried out by Pecoud et al. [14]. Thus it is clear that for effective administration of zinc in therapeutic situations to achieve pharmacological tissue levels, zinc should not be given with most foods. We recommend zinc ingestion at least 1 hr before or at least 1 hr after meals.

Even if food is avoided, only about 10% of orally administered zinc is absorbed. While the factors influencing zinc absorption have received some study, the exact mechanisms by which zinc absorption is limited in this manner are not understood.

3. ORAL REGIMENS

As discussed above, the traditional regimen of zinc is 220 mg of zinc sulfate (50 mg of elemental zinc) administered three times a day with meals. However, as we have discussed in the previous section, zinc administered along with food is not well absorbed, and thus we recommend that zinc be given 1 hr before or after meals. Also, the elevation of plasma zinc from an oral dose of zinc lasts only about 5 hr [13]. As a result, if one desires continuous elevation in plasma zinc, it is probably necessary to administer the zinc more often than three times a day. Thus, we administer zinc in five or six divided doses per 24 hr. The other factor which should be mentioned is that zinc sulfate is somewhat irritating to the gastrointestinal tract, particularly if administered without food. It is our belief that as with ferrous sulfate, a significant component of the irritation is due to the sulfate moiety. In preliminary studies we have shown that zinc acetate is as well absorbed as zinc sulfate, and appears to be somewhat better absorbed than zinc carbonate [13]. From additional preliminary work in human subjects, it appears that zinc acetate is better tolerated than zinc sulfate. This has not been confirmed by a controlled study as yet, but we use zinc acetate routinely.

Thus, the therapeutic regimen we are currently using is 25 mg of elemental zinc, administered as the acetate, five or six times a day, avoiding food 1 hr before and 1 hr after each dose. For many of our patients, the schedule we employ is to take 25 mg of zinc at 7:00 and 11:00 a.m. and 3:00, 7:00, and 11:00 p.m. For most patients, the 3:00 a.m. dose is skipped, and the patients simply take an additional 25 mg of zinc with their 11:00 p.m. or 7:00 a.m. dose.

4. DISEASES WHICH MAY RESPOND TO ZINC THERAPY

4.1. Sickle Cell Anemia

Early studies with zinc showed that zinc would improve the ability of partially deoxygenated sickle cells to go through small pores of Nucleopore filters [7]. Of course, deoxygenation causes sickle cells to begin sickling and decreases their deformability, making it more difficult for them to go through small pores. It was shown that zinc had this beneficial effect on filterability at concentrations too low to be interacting directly with hemoglobin. That is, less than one molecule of zinc per 100 molecules of hemoglobin tetramer improved sickle cell filterability. Since it was known that sickle cells develop a membrane lesion which decreases their deformability, it was proposed that zinc had some type of beneficial effect on the membrane [7].

Based upon these in vitro empiric observations, and also because of the low toxicity of zinc, oral zinc was administered in the regimen described earlier to sickle cell anemia patients. Slight improvement in hemoglobin levels and in chromium-51 red cell survivals were noted. Most striking, however, was the reduction in the count of irreversibly sickled cells (ISCs) [8]. The ISC is a membrane-damaged cell. These cells will not return to a normal configuration no matter how long they are oxygenated. Evidence of damage to the membrane of these cells can be obtained when, after lysing the cells and allowing the release of soluble hemoglobin, the remaining membrane remains in the abnormal shape [17]. Further, most of the lipid of the membrane can be removed, and the remaining protein cytoskeleton will remain in the ISC configuration [18]. Thus, somehow the protein cytoskeleton of the membrane has become locked into abnormal configuration. An example of a blood film showing numerous ISCs is shown in Fig. 2.

In order to evaluate the effect of therapy on the ISC count it is necessary to develop specific criteria for ISCs, practice counting coded slides until counts are reproducible, and then use coded slides throughout the study. If one does this, ISC counts are quite

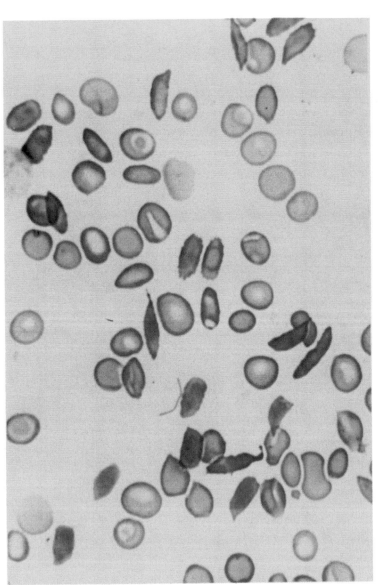

FIG. 2. This blood film from a sickle cell anemia patient has been fully oxygenated, allowing reversibly sickled cells to return to a normal round shape. However, numerous elongated and abnormally shaped cells remain. These are the irreversibly sickled cells (ISCs), which have permanent membrane damage preventing their return to normal. The number of these cells in the circulation can be significantly reduced by oral zinc administration to the patient.

reproducible and show patient specificity. That is, while some sickle cell anemia patients show counts in the 40% range and others in the 10% range, a given patient shows variation only within fairly narrow limits around a particular value [8,19].

With our oral zinc regimen we were able to reduce the mean ISC count in 12 patients from 28 to 18.6% (p < 0.001) [8]. Patients with the highest ISC counts showed the greatest effect. Because there is mild hematologic improvement with zinc, we believe that the effect of zinc on ISCs is to prevent their formation rather than to increase their destruction. The possible mechanism of the beneficial effect of zinc on the membrane lesion in sickle cell anemia will be discussed in the next section.

Patients with sickle cell anemia, in addition to having anemia, also have difficulties stemming from vasoocclusion leading to accumulated organ damage and to episodes of pain called sickle cell pain crises. The effects of zinc on prevention of organ damage in sickle cell anemia is unknown. It might take a very-long-term clinical trial to evaluate benefits of this type. However, it can be noted that certain kinds of organ damage seem to be positively correlated with the ISC count. For example, the severity of vascular damage that can be seen in the conjunctival blood vessels of the eyes is positively correlated with the ISC count [20]. It has been shown that the spleen is destroyed by autoinfarction at a younger age in patients with a higher ISC count [21], and the loss of renal function resulting in hyposthenuria also occurs earlier in patients with higher ISC counts [22]. This may all mean that the reduction in ISC counts obtained with zinc therapy indicates benefit in terms of prophylaxis of organ damage, but this is unknown at this point.

Many patients with sickle cell anemia have frequent pain crises. These are episodes which may last a few hours to more than 2 weeks, and may be so severe as to require hospitalization. Their exact basis is unknown, but they are presumed to result from vaso-occlusive events. In uncontrolled, relatively short-term studies, zinc has a substantial prophylactic effect in preventing such pain crises. However, this could be due to placebo effect. In a 3-year

double-blind study in nine patients, an effect of zinc on reducing
pain crisis could not be shown [23]. However, we have suspected
that compliance with the difficult regimen of zinc administration
described in Sec. 3 probably was not good during this long study.

4.2. Rheumatoid Arthritis

In 1976 Simkin [15,24] reported a beneficial effect of administration of 220 mg of zinc sulfate, given three times a day with meals,
to patients with rheumatoid arthritis. This was a short-term study
designed simply to test for an effect and not to evaluate long-term
management. The rationale for this study was multiple, including:
(1) rheumatoid arthritis patients may often be zinc-deficient [15];
(2) rheumatoid arthritis patients have low serum histidine levels,
and histidine and zinc metabolism and absorption may be linked [15];
(3) rheumatoid arthritis patients derive benefit from treatment with
the chelating agent, penicillamine, which may enhance zinc absorption [15]; and (4) zinc has been found to have an anti-inflammatory
effect [15]. Simkin's results [15] showed a significant improvement
in joint swelling, tenderness, and morning stiffness. To our knowledge, this promising beginning has not been followed up with additional clinical studies.

We have carried out an animal study which is related in that it
deals with another autoimmune disease, an animal model of systemic
lupus erythematosus (SLE). In this mouse model, all F_1 progency of
an NZB/NBW cross developed SLE. In our study, administration of zinc
over periods of months produced significant protection against the
renal lesion in these animals as compared to controls [25].

Based upon the results in humans in rheumatoid arthritis, and
in the SLE mouse model, it appears that zinc may offer significant
therapeutic effects in certain autoimmune disorders. It appears
that the effects of zinc on these inflammatory diseases is not on
the antibody-producing cells or on antibody levels, because, for
instance, in the lupus mouse model, zinc therapy had no effect on the

levels of antinuclear antibody [25]. Our current rationale for the
beneficial effect of zinc relates to inhibition of inflammatory
cells, a rationale which will be discussed in the next section. If
zinc does have significant anti-inflammatory effects, then a variety
of inflammatory diseases might be successfully treated with zinc.
Thus, diseases such as ulcerative colitis, regional ileitis, certain
rheumatologic diseases, and any disease in which an inflammatory
response is an important component should be considered.

5.   MECHANISM OF ACTION

5.1.   Sickle Cell Anemia

Our early work on the mechanism of action of zinc on the membrane
lesion of the sickle cell centered on the hypothesis that zinc was a
calcium antagonist [7,9]. A considerable amount of data has accumulated involving a variety of cell types showing that zinc and calcium
have reciprocal or antagonistic effects on cells. This is true of
the mast cell in which histamine release is calcium-dependent [26,
27] and inhibited by zinc [28,29]; it is true of the platelet in
which release and aggregation are calcium-dependent [30,31] and inhibited by zinc [32]; and it is true of the neutrophil and of the
macrophage in which activation requires calcium [33] and is inhibited by zinc [34,35]. In the red cell, zinc displaces calcium
from the membrane [7], decreases calcium-stimulated binding of hemoglobin to the inner surface of the membrane [36], inhibits calcium-
stimulated echinocyte formation [37], and inhibits calcium-induced
red cell shrinkage [38].

This zinc-calcium antagonism requires a molecular explanation.
Our present hypothesis is that the molecular mechanism involves
reciprocal effects on the protein calmodulin [11,12]. Calmodulin is
a 17,000 molecular weight protein present in most cells, including
the erythrocyte [39-42]. It has four calcium binding sites and when
activated by an increased intracellular calcium, has a capacity to

activate multiple proteins. The exact spectrum of action of calcium-
activated calmodulin depends upon the cell type. In the erythrocyte,
one known function is activation of calcium-ATPase, the calcium pump
of the erythrocyte [41]. However, it is believed that it probably
has multiple actions even in the erythrocyte. Since during sickling
an influx of calcium occurs [10], it is possible that calmodulin is
excessively activated and causes damage to the membrane. We have
postulated that zinc owes its anticalcium action and possibly its
antisickling action to inhibition of calmodulin function [11,12].
We have shown that the activation of three different enzymes by calmodulin is inhibited by zinc. These are calcium-APTase, adenyl
cyclase, and phosphodiesterase.

That calmodulin is involved in many of the cellular reactions
inhibited by zinc is further supported by the inhibition of many
such cellular reactions by the phenothiazine, trifluoperazine. Trifluoperazine (and other phenothiazines) bind to and inhibit calmodulin [43,44] and have been shown to inhibit platelet aggregation [45]
and neutrophil activation [46].

There is an interesting connection between calmodulin-inhibiting properties of drugs and their ability to cause membrane expansion. Membrane expansion was studied most widely by Seeman in the
1960s and early 1970s [47]. According to Seeman, many drugs, including the phenothiazines, increase the surface area of the red
cell membrane, often 10-fold more than can be accounted for by
direct incorporation of the drug into the membrane. This property
can be detected by protection afforded the cell against osmotic
lysis. That is, when red cells are exposed to hypotonic conditions,
the surface-area-to-volume ratio determines the amount of lysis with
a given hypotonic salt concentration. Seeman showed that the phenothiazines and many other drugs protect against osmotic lysis and are
therefore membrane expanders. The drugs which have been previously
demonstrated to be calmodulin inhibitors, including the phenothiazines, the tricyclic antidepressants, and the butyrphenones [48],
are all membrane expanders [47].

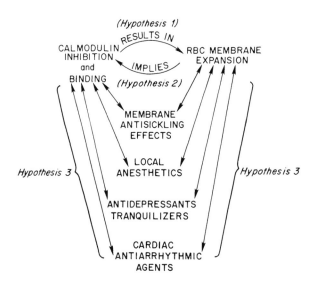

FIG. 3. Hypothesized relationships between calmodulin inhibition, red cell membrane expansion, and certain pharmacological effects. Zinc fits into this scheme as an inhibitor of calmodulin function, as a membrane expander, and as a membrane-antisickling agent. (From Ref. 50.)

These associations have led us to the concept proposed in Fig. 3--that is, that calmodulin inhibition and membrane expansion are associated properties of classes of drugs with certain pharmacological actions. The types of pharmacological actions are indicated in the lower part of Fig. 3. To test out the validity of these proposed associations we have carried out some specific experiments. First, we have found that the local anesthetics, which were shown by Seeman [47] to be membrane expanders, are inhibitors of calmodulin activation of calcium-ATPase and phosphodiesterase [49,50]. Thus, these drugs are inhibitors of calmodulin function. Second, we have shown that zinc, which, as we have already discussed, inhibits calmodulin functions, is a membrane expander [49,50]. Zinc, procaine, and the phenothiazines, all of which expand membranes and inhibit calmodulin, have membrane-antisickling properties [12,51,52]. This suggests that antisickling drugs which act on the membrane will tend to come from membrane expander/calmodulin inhibitor classes of

drugs. To test this concept, we studied cetiedil, which has recently been shown to have membrane-antisickling properties [53,54].
We have shown that cetiedil both expands membranes and inhibits calmodulin [49,50]. Further, we have noted that certain cardiac antiarrhythmic agents, such as propranolol and procainamide, were shown
to be membrane expanders by Seeman [47]. We have found that these
drugs inhibit calmodulin activation of calcium-ATPase and phosphodiesterase. Because quinidine is a cardiac antiarrhythmic agent, we
tested it for inhibition of calmodulin activation of these enzymes
and for membrane expansion. It does both [49,50]. Thus, a pattern
is emerging whereby there is an association between calmodulin inhibition, membrane expansion, and certain pharmacological properties
of drugs. Zinc seems to fit into this pattern of being a membrane
expander/calmodulin inhibitor, and of the pharmacological properties
it possesses at least membrane-antisickling effects.

5.2. Anti-inflammatory Effects

Chvapil et al. must be credited with being the first to call attention to the anti-inflammatory effects of zinc [55]. Chvapil, Zukoski,
and colleagues have shown that zinc inhibits granulocyte junction
[35] and macrophage function [56] and decreases the inflammatory
response in animals to agents stimulating inflammation, such as
silica particles in the rat [57]. Thus it seems reasonable to suggest that the anti-inflammatory effects of zinc, for example, in
rheumatoid arthritis [24] or mouse SLE [25], may involve inhibition
of inflammatory cells.

Regarding the molecular mechanism of the anti-inflammatory effects of zinc, Chvapil and colleagues have proposed a variety of
possible effects [34,57], including "membrane stabilization," effects
on membrane reactive groups such as sulfhydryls, effects on enzymes
controlling membrane receptors, displacement of calcium from intracellular components such as microfilaments or microtubules, complexing NADPH, reacting with superoxide dismutase, etc. The variety of

proposed possible mechanisms indicates that no central theory of molecular mechanism of action has yet been adopted.

We now hypothesize that the molecular mechanism of zinc's anti-inflammatory action is based upon the properties discussed in the previous section, involving membrane expansion and inhibition of calmodulin function. In this mechanism, the calcium-dependent functions of inflammatory cells, including activation of calmodulin, are inhibited by zinc.

6. TOXICITY

Zinc is a very nontoxic drug. Given orally, it does occasionally produce some irritation of the gastric mucosa. This can usually be handled by administering the zinc with something which does not block its absorption, such as meat or soft drinks. We have switched from zinc sulfate to zinc acetate to eliminate the gastric irritation of the sulfate moiety, and find zinc acetate to be reasonably well tolerated.

A more significant toxicity of zinc when given in pharmacological doses is copper deficiency [58,59]. Zinc-induced copper deficiency has been seen in animal studies and thus is well established [60]. We observe zinc-induced copper deficiency in our sickle cell anemia patients, in that after several months of treatment most of them tend to develop chemical copper deficiency. This is manifested by a reduction in ceruloplasmin levels. If this is allowed to progress, a hypochromic-microcytic red cell picture and neutropenia emerge. This is easily corrected by administration of 0.5 mg of copper as copper sulfate per day. The mechanism of zinc-induced copper deficiency appears to be competition for absorption in the gastrointestinal tract.

There appear to be no storage disease of zinc and very little other recognized toxicity. One would imagine an increased susceptibility to infection might occur in view of inhibition by zinc of neutrophils and macrophages, but this has not yet been reported.

7. THE FUTURE

In terms of the mechanism of zinc action, it is important to evaluate the role of zinc as an inhibitor of calmodulin. One of the first things that needs to be done is to evaluate the actual binding of zinc to calmodulin. If it is found that zinc does not bind to calmodulin at concentrations that can be achieved pharmacologically, then the next step is to evaluate whether all or most of the target proteins of calmodulin bind zinc. Included in this analysis should be the possible membrane binding sites of calmodulin. If zinc does not owe its membrane action to inhibition of calmodulin or its target proteins, other membrane mechanisms of action must be sought. Perhaps the membrane expansion effect of zinc could be a clue.

In terms of sickle cell anemia, a way must be found, if zinc is to be developed for optimal clinical effects, to increase the plasma levels of zinc. One way to accomplish this would be to develop a prodrug approach, that is, to develop a mechanism to enhance the oral absorption of zinc. As we indicated earlier, only about 10% of administered zinc is absorbed. Another approach is to develop an effective method for the parenteral administration of zinc. The evidence that we have from animal model studies indicates that if plasma zinc levels could be increased to perhaps 250-300 µg%, zinc therapy would be much more effective [61]. An additional approach is to use the knowledge about the probable mechanism of zinc action, namely, calmodulin inhibition, to combine zinc with other drugs which have calmodulin-inhibiting properties. Our present work involves a combination of zinc and phenothiazines as a treatment for sickle cell anemia.

ACKNOWLEDGMENTS

This work was supported by grants from the Herrick, Sage, and Ervin Foundations.

REFERENCES

1. W. J. Pories, E. G. Mansour, F. R. Plecha, A. Flynn, and W. H. Strain, in *Trace Elements in Human Health and Disease* (A. S. Prasad and D. Oberleas, eds.), Academic Press, New York, 1976, pp. 115 ff.

2. W. H. Strain, A. M. Dutton, H. B. Heyer, and G. H. Ramsey, *University of Rochester Reports,* p. 18 (1953).

3. W. E. C. Wacker, in *Trace Elements in Human Health and Disease* (A. S. Prasad and D. Oberleas, eds.), Academic Press, New York, 1976, pp. 107 ff.

4. W. J. Pories, J. H. Henzel, C. G. Rob, and W. H. Strain, *Lancet, 1,* 121 (1967).

5. W. J. Pories, J. H. Henzel, C. G. Rob, and W. H. Strain, *Ann. Surg., 165,* 432 (1967).

6. G. R. Serjeant, R. E. Galloway, and M. C. Gueri, *Lancet, ii,* 891 (1970).

7. G. J. Brewer and F. J. Oelshlegel, Jr., *Biochem. Biophys. Res. Commun., 58,* 854 (1974).

8. G. J. Brewer, L. F. Brewer, and A. S. Prasad, *J. Lab. Clin. Med., 90,* 549 (1977).

9. G. J. Brewer, F. J. Oelshlegel, Jr., and A. S. Prasad, in *Erythrocyte Structure and Function* (G. J. Brewer, ed.), Alan R. Liss, New York, 1975, pp. 417 ff.

10. J. W. Eaton, T. D. Skelton, J. S. Swofford, C. F. Kolpin, and H. S. Jacob, *Nature, 246,* 105 (1973).

11. G. J. Brewer, J. C. Aster, C. A. Knutsen, and W. C. Kruckeberg, *Amer. J. Hemat., 7,* 53 (1979).

12. G. J. Brewer, *Amer. J. Hemat., 8,* 231 (1980).

13. F. J. Oelshlegel, Jr., and G. J. Brewer, in *Zinc Metabolism: Current Aspects in Health and Disease* (G. J. Brewer and A. S. Prasad, eds.), Alan R. Liss, New York, 1977, pp. 299 ff.

14. A. Pecoud, P. Donzel, and J. L. Schelling, *Clin. Pharmacol. Ther., 17,* 469 (1975).

15. P. A. Simkin, in *Zinc Metabolism: Current Aspects in Health and Disease* (G. J. Brewer and A. S. Prasad, eds.), Alan R. Liss, New York, 1977, pp. 343 ff.

16. A. S. Prasad, in *Trace Elements in Human Health and Disease* (A. S. Prasad, ed.), Academic Press, New York, 1976, pp. 1 ff.

17. J. F. Bertles and J. Dobler, *Blood, 33,* 884 (1969).

18. S. E. Lux, K. M. John, and M. J. Karnovsky, *J. Clin. Invest., 58,* 955 (1976).

19. G. R. Serjeant, B. E. Serjeant, and P. F. Milner, *Brit. J. Haematol.*, *17*, 527 (1969).

20. G. R. Serjeant, B. E. Serjeant, and P. I. Condon, *J. Amer. Med. Ass.*, *219*, 1428 (1972).

21. G. R. Serjeant, *Brit. J. Haematol.*, *19*, 635, 1970.

22. R. Warrier, S. Sarnik, and G. J. Brewer, *Pediatr. Res.*, *13*, 443 (1979).

23. G. J. Brewer, unpublished results.

24. P. A. Simkin, *Lancet, ii*, 539 (1976).

25. P. Kretzschmar, G. J. Brewer, and S. E. Walker, submitted for publication.

26. S. A. Patkar and B. Diamant, *Agents Actions*, *4*, 200 (1974).

27. N. Grosman and B. Diamant, *Agents Actions*, *4*, 198 (1974).

28. B. Högberg and B. Uvnäs, *Acta Physiol. Scand.*, *48*, 133 (1960).

29. W. Kazimierczak and C. Maslinski, *Agents Actions*, *4*, 320 (1970).

30. A. Pletscher, M. DaPrada, and J. P. Tranzer, *Experientia* (Basel), *24*, 1202 (1968).

31. M. G. Ardlie, E. E. Nishizawa, and M. Guciione, *Fed. Proc.*, *20*, 423 (1970).

32. M. Chvapil, P. L. Weldy, L. Stankova, D. S. Clark, and C. F. Zukoski, *Life Sci.*, *16*, 561 (1975).

33. T. P. Stossel, *J. Cell. Biol.*, *58*, 346 (1973).

34. M. Chvapil, *Med. Clin. North Amer.*, *60*, 799 (1976).

35. M. Chvapil, L. Stankova, C. Zukoski IV, and C. Zukoski III, *J. Lab. Clin. Med.*, *89*, 135 (1977).

36. S. Dash, G. J. Brewer, and F. J. Oelshlegel, Jr., *Nature, 250*, 251 (1974).

37. W. C. Kruckeberg, F. J. Oelshlegel, Jr., S. H. Shore, P. E. Smouse, and G. J. Brewer, *Res. Exp. Med.*, *170*, 149 (1977).

38. J. W. Eaton, E. Berger, J. G. White, and H. S. Jacob, in *Zinc Metabolism: Current Aspects in Health and Disease*, Alan R. Liss, New York, 1976, pp. 275 ff.

39. W. Y. Cheung, *Biophys. Res. Commun.*, *38*, 533 (1970).

40. S. Kakiuchi, R. Yamazaki, and H. Nakajima, *Proc. Jap. Acad.*, *46*, 587 (1970).

41. G. H. Bond and D. L. Clough, *Biochim. Biophys. Acta*, *323*, 592 (1973).

42. W. Y. Cheung, T. J. Lynch, and R. W. Wallace, in *Advances in Cyclic Nucleotide Research* (W. J. Beorge and L. F. Ignarro, eds.), Raven Press, New York, 1978, pp. 233 ff.

43. R. M. Levin and B. Weiss, *Molec. Pharmacol., 13,* 690 (1977).

44. R. M. Levin and B. Weiss, *Biochim. Biophys. Acta, 540,* 197 (1978).

45. W. C. Kruckeberg and G. J. Brewer, *Clin. Res., 25,* 612A (1977).

46. J. Silva, R. Amirkanian, J. Aster, and G. Brewer, *Clin. Res., 28,* 360A (1980).

47. P. Seeman, *Pharmacol. Rev., 24,* 583 (1972).

48. R. M. Levin and B. Weiss, *J. Pharmacol. Exp. Ther., 208,* 454 (1979).

49. U. L. Bereza and G. J. Brewer, *Clin. Res., 28,* 759A (1980).

50. G. J. Brewer, U. Bereza, I. Mizukami, J. C. Aster, and L. F. Brewer, in *The Red Cell--Fifth Ann Arbor Conference* (G. J. Brewer, ed.), Alan R. Liss, New York, 1981.

51. R. Baker, D. Powars, and L. J. Haywood, in *Erythrocyte Structure and Function* (G. J. Brewer, ed.), Alan R. Liss, New York, 1977, pp. 453 ff.

52. R. A. Lewis and F. N. Gyang, *Arch. Int. Pharmacodyn. Ther., 153,* 158 (1965).

53. L. J. Benjamin, G. Kokkini, and C. M. Peterson, *Blood, 55,* 265 (1980).

54. T. Asakura, S. T. Ohnishi, K. Adachi, M. Ozguc, K. Hashimoto, M. Singer, M. Russel, and E. Schwartz, *Proc. Nat. Acad. Sci. U.S., 77,* 2955 (1980).

55. M. Chvapil, S. L. Elias, J. N. Ryan, and C. F. Zukoski, in *International Review of Neurobiology,* Suppl. 1, Academic Press, New York, 1972, pp. 105 ff.

56. C. F. Zukoski, M. Chvapil, E. C. Carlson, B. Hattler, and J. C. Ludwig, *J. Res., 16,* 6a (1974).

57. M. Chvapil, L. Stankova, P. Weldy, D. Bernhard, J. Campell, E. C. Carlson, T. Cox, J. Peacock, Z. Bartos, and C. Zukoski, in *Zinc Metabolism: Current Aspects in Health and Disease* (G. J. Brewer and A. S. Prasad, eds.), Alan R. Liss, New York, 1977, pp. 103 ff.

58. G. J. Brewer, E. B. Schoomaker, D. A. Leichtman, W. C. Kruckeberg, L. F. Brewer, and N. Meyers, in *Zinc Metabolism: Current Aspects in Health and Disease* (G. J. Brewer and A. S. Prasad, eds.), Alan R. Liss, New York, 1977, pp. 241 ff.

59. A. S. Prasad, G. J. Brewer, E. B. Schoomaker, and P. Rabbani, *J. Amer. Med. Ass., 240,* 2166 (1978).

60. C. H. Hill, in *Trace Elements in Human Health and Disease* (A. S. Prasad, ed.), Academic Press, New York, 1976, pp. 281 ff.

61. E. B. Schoomaker and G. J. Brewer, in *The Red Cell* (G. J. Brewer, ed.), Alan R. Liss, New York, 1978, pp. 177 ff.

Chapter 4

THE ANTI-INFLAMMATORY ACTIVITIES
OF COPPER COMPLEXES

John R. J. Sorenson
Department of Biopharmaceutical Sciences
College of Pharmacy and Department of
Pharmacology, College of Medicine
University of Arkansas for Medical Sciences
Little Rock, Arkansas

1. INTRODUCTION

Inflammation is an important response to tissue injury due to any
cause. The importance of this multifaceted process is best appreciated as the beginning of the tissue repair process, which is required to reestablish normal function. Persistent inflammation in
the absence of tissue repair and the resultant lack of normal function is recognized as chronic inflammation.

    An "anti-inflammatory" agent that inhibits some facet of the
inflammatory process and, as a result, the repair process would not
be expected to reestablish normal function. Alternatively, an anti-
inflammatory agent which facilitates the repair process would be expected to reestablish normal function. In retrospect, many anti-
inflammatory agents have been developed to inhibit some component of
the inflammatory process without correcting the cause of the disease
or promoting tissue repair. Enough mechanistic evidence is available

to support the claim that copper complexes do not fit into this category of anti-inflammatory agents. They have been demonstrated to promote tissue repair processes. As a result, these anti-inflammatory compounds merit serious consideration for the treatment of inflammatory diseases. The mechanistic data which account for the effectiveness of these copper-containing compounds has also provided a better understanding of inflammatory diseases.

As it happens, many diseases which are not generally recognized as inflammatory diseases do have an inflammatory component which requires tissue repair to attain normal function. Appreciation for this point makes it easier to understand why anti-inflammatory agents which promote tissue repair are also effective in treating or preventing these other diseases. It is recognized that chemical or immunologically induced irritations cause local or systemic inflammations and that these inflammations can be prevented or treated with an anti-inflammatory agent which promotes tissue repair. Recognizing that gastric ulcers are, in part, inflammatory diseases of the stomach accounts for the effectiveness of an anti-inflammatory agent which promotes tissue repair processes in the prevention or treatment of gastric ulcers. Recognizing that cancers are in part inflammatory diseases of the sites of cancerous growth accounts for the effectiveness of an anti-inflammatory agent which promotes tissue repair processes in the prevention or treatment of cancers. Recognizing that seizures may be, in part, an inflammatory disease of the central nervous system accounts for the effectiveness of an anti-inflammatory agent which promotes tissue repair processes in the prevention or treatment of seizures. Recognizing that still other diseases or physiological states are in part inflammatory may account for other pharmacological activities of anti-inflammatory agents which promote tissue repair processes.

The following evidence is provided in support of the anti-inflammatory activities of copper complexes and their potential as antiarthritic, antiulcer, anticancer, and antiepileptic drugs. The anti-inflammatory, antiulcer, anticancer, and anticonvulsant

activities of copper complexes support the hypothesis that the corresponding diseases are, in part, inflammatory diseases and emphasize the need for more research in the area of essential metal
metabolism and inflammation.

2. NORMAL PHYSIOLOGY OF COPPER COMPLEXES

Because of the affinity of ionic copper for coordinating ligands,
the concentration of free ionic copper is likely to be too small to
measure in any biological system. The concentration of ionic copper
in blood plasma has been estimated to be approximately 10^{-20} mol/ml
[1]. As a result, in considering the physiology of copper it must
be borne in mind that complexes are, by far, the predominant form of
this element. Copper complexes are ingested, absorbed, and transported in blood plasma. The copper in these complexes is stored or
utilized in some copper-dependent process following ligand exchange,
or excreted as a complex. At each stage of this metabolic sequence,
the complexes may undergo ligand exchange to form other complexes.

   Copper is known to be essential because it, like an essential
amino acid, an essential fatty acid, or an essential cofactor
(vitamin), is required for normal metabolism but is not synthesized
in the body. A number of excellent reviews on the physiology of
copper have been published by authorities in this area [2-9].

   The adult body contains between 1.4 and 2.1 mg of copper per
kg body weight [10]. The infant body contains 3 times this amount
of copper, which is consistent with the fact that their metabolic
needs are much greater than adults. All tissues of the body need
copper for normal metabolism, but some tissues have greater metabolic needs than others. As a result, these tissues contain more
copper. The amounts of copper found in various body tissues and
fluids are shown in Table 1.

   The amount of copper in each tissue correlates with the number
and kind of metabolic processes requiring copper in that tissue. In
this regard, it is of interest to point out that the brain and heart

TABLE 1

Mean Concentration of Cu in Tissues and Fluids
(μg/g of tissue ash or as shown)

Adrenal	210	Milk	
Aorta	97	Colostrum	0.35-0.50 μg/ml
Bile	547	Mature	0.20-0.50 μg/ml
Blood (total)	1.01 μg/ml	Muscle	85
Erythrocytes	0.98 μg/ml	Nails	23 μg/g
Plasma	1.12 μg/ml	Omentum	190
Serum	1.19 μg/ml	Ovary	130
Bone	25 μg/g	Pancreas	150
Brain	370	Pancreatic fluid	105
Breast	6 μg/g	Placenta	4 μg/g
Cerebrospinal fluid	0.22 μg/g	Prostate	110
Diaphragm	150	Saliva	0.08 μg/ml
Esophagus	140	Skin	120
Gallbladder	750	Spleen	93
Hair	19 μg/g	Stomach	230
Heart	350	Sweat	0.55 μg/ml
Intestine		Testes	95
Cecum	220	Thymus	4 μg/g
Duodenum	300	Thyroid	100
Ileum	· 280	Tongue	4.6 μg/g
Jejunum	250	Tooth	
Rectum	180	Dentine	2 μg/g
Signoid colon	230	Enamel	10 μg/g
Kidney	270	Trachea	65
Larynx	59	Urinary bladder	120
Liver	680	Urine	0.04 μg/ml
Lung	130	Uterus	110
Lymph node	60		

Source: Data from Refs. 11 and 12.

contain more copper than all other tissues except the liver, which
is the major copper storage organ. Because the gallbladder and bile
serve as the major excretory route of copper, they also contain
large amounts. Organs with the next most abundant content in decreasing order are the stomach, intestine, and adrenal glands. The
remaining tissues have lesser amounts of copper because of their
relatively lower metabolic activity, but it is just as important for
normal metabolism in these tissues as it is in all others.

Although bile is the major excretory vehicle for excess copper, significant but lesser amounts are lost via the hair, outer layer of skin, finger- and toenails, sweat, and urine as end products of metabolism. These losses point out the need for a compensating daily intake *and absorption* to replenish this essential element.

Ingested copper follows the pathway presented in Fig. 1. One of a large number of possible copper complexes that might be found

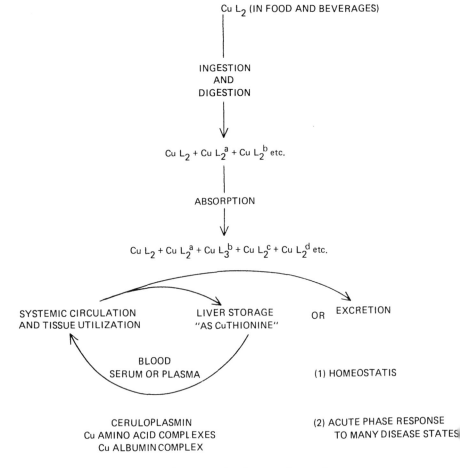

FIG. 1.   Distribution of copper in ingested complexes.

in foods and/or beverages, following ingestion and digestion, would give rise to the formation of other copper complexes as a result of exchange with ligands in the digestive tract, but some of the originally ingested complex may remain intact. Additional complexes may be formed by ligand exchange following adsorption, but some of the original complex may still remain. These copper complexes then undergo systemic circulation to all tissues. As a result of this circulation, the copper in these complexes is either (1) utilized by tissues following ligand exchange with apoenzymes and apoproteins to form metalloenzymes and metalloproteins; (2) stored in the liver following exchange with thionine to form copper-thionine; or (3) excreted via bile or urine.

As shown in Fig. 1, stored copper may be released by the liver into blood to meet normal metabolic needs (1). This homeostatic release of copper from the liver meets the normal physiologic requirements of body tissues. Larger quantities of copper are released as part of an acute-phase response in many disease states (2). This response may result in a two- to threefold increase in blood copper concentration which reflects a greater physiologic need.

Three forms of copper are released from the liver. The majority is released as ceruloplasmin, an enzyme containing six to eight atoms of tightly bound copper [9]. The two remaining forms of copper complexes are less firmly bound amino acid copper complexes and a copper albumin complex [13,14]. Normal amounts of these blood components are synthesized and released from the liver to meet normal physiologic needs, but much greater amounts are released to meet increased requirements in response to many disease states.

To date, it has been established that copper is required for hemoglobin synthesis, growth, keratinization, pigmentation, bone formation, reproduction, fertility, development, and function of the central and peripheral nervous systems, cardiac function, cellular respiration, nerve function, extracellular connective tissue formation, vascularization, mental and behavioral development, sightedness, and regulation of monoamine concentrations [10]. These roles

for copper are based upon the requirement of coordinated copper at
the active site of the following copper-dependent enzymes: ceruloplasmin, tyrosinase, lysyl oxidase, ceramide galactosyl transferase;
cytochrome-c oxidase, dopamine-β-hydroxylase, pyridoxal-requiring
monoamine oxidases, and a superoxide dismutase [4,9]. In addition,
copper-dependent processes appear to be required for the modulation
of prostaglandin syntheses [15], lysosomal membrane stabilization
[15], and the modulation of histaminic activity [16].

Copper has received considerable attention with regard to its
presence in normal blood plasma and serum components. Perusal of
this literature revealed that total serum copper concentrations were
most commonly determined and reported through the years to be between
94 and 133 μg/100 ml [17]. Variation in these values is in part due
to geographic location, which may affect dietary intake, but sample
contamination and analytical error have been cited to account for
marked deviations from the currently accepted norm of 110 $\pm$ 12 μg/
100 ml for males and 123 $\pm$ 16 μg/100 ml for females [10]. Larger
than usual mean copper concentrations may also be the result of including in a "normal" population those individuals who have an infection or inflammation of one sort or another, since these diseases
are associated with elevated copper concentrations. Sex-related
variation in serum copper has been attributed to observations that
estrogens elevate serum copper values [18-21].

3. ALTERED PHYSIOLOGY OF COPPER COMPLEXES IN DISEASE WITH INFLAMMATORY COMPONENTS

3.1. Rheumatoid Arthritis

A review of the literature published over the last 40 years concerning studies of copper concentrations in tissues from individuals with
rheumatoid arthritis has revealed that concentrations are altered in
this disease state [17]. Compared to normal healthy individuals,
patients with rheumatoid arthritis have higher mean serum or plasma
copper concentrations which are directly related to disease severity

or activity as measured by increased body and local temperatures, immobility, duration of disease, pain, edema, and erythrocyte sedimentation rate, as well as diminished strength and decreased hemoglobin values. Small sex-related differences in normal individuals are obscured by marked increases found for both male and female patients. Copper concentrations increase in association with the onset and persistence of active disease but return to normal with remission.

It is now known from animal studies that the rise in serum copper is accompanied by a fall in liver copper concentrations [22-24]. Since serum copper-containing components are synthesized in the liver and appear in serum after the onset of disease, it seems reasonable to suggest that this is a physiologic response to arthritis, which facilitates remission. Alternatively, a lack of this response was suggested to result in chronic or persistent disease [17]. This view is consistent with the observations that blood serum or plasma levels of these same copper-containing components increase in animals in response to inflammations [17,24-26] and that these inflammations are more severe inflammatory diseases in animals with established copper deficiency [27-30]. All of the above support the role of the increased copper-containing components in blood as "putative modulators" of inflammation [31].

The low or normal serum copper concentrations found for some arthritic patients deserve some comment. These values may be the result of a failure of this aspect of the physiologic response due to depletion of liver copper stores. Depletion may be due to increased turnover resulting in copper excretion [32,33] and failure to replete these stores because of either loss of appetite or inadequate diet. Failure of serum copper to increase maximally as a physiologic response to disease could lead to chronic disease.

In summary, the results reported since 1937 [17] have increased our understanding of essential metal metabolism associated with rheumatoid arthritis. Marked increases in serum or plasma copper were confirmed in patients with active disease. The increased rate of synthesis and the accelerated turnover rate of ceruloplasmin were found to be directly related to disease activity. Copper was also found to increase in synovial fluid of rheumatoid arthritis patients and it was suggested that ceruloplasmin, which accounted for most of the rise in synovial fluid copper, increased with increasing duration of disease. Accelerated rates of ceruloplasmin synthesis and turnover, marked increases in serum copper, and accumulation of ceruloplasmin copper in synovial fluid and an increase in C-reactive protein [17,34] can all be interpreted as part of a multifaceted physiologic response to inflammatory disease.

With regard to altered concentrations of copper in arthritis, it is of special interest to note that a number of copper-dependent enzymes are required for the repair of inflammed tissues. These include cytochrome-c oxidase, ceruloplasmin, lysyl oxidase, superoxide dismutase, and possibly tyrosinase and dopamine-β-hydroxylase. In addition, the copper-dependent modulating process involved in prostaglandin syntheses, lysosomal membrane stabilization, and histamine mediation of vascular and secretory activities may also be important in tissue repair processes [35]. Additional evidence in support of a role for copper in the repair of arthritic tissues is provided in Secs. 4.1, 5, and 6 of this chapter.

3.2. Gastric Ulcers

To date there are no published studies of serum or plasma copper concentrations in patients with gastric ulcers. However, since serum or plasma copper is known to increase as an acute-phase response in various disease states [10], it is likely that the copper concentrations increases in this disease state as well. This conclusion is supported by the recognition that gastric ulcers are inflammatory disease of the gastric mucosa and submucosa and that gastric ulcers are often a component of arthritic disease [36-37].

Because gastric ulcers are an extraarticular inflammatory disease, the important copper-dependent enzymes listed in Sec. 3.1

as required for the repair of arthritic inflammatory disease are
also considered important for the repair of gastric ulcers. Evidence in support of this hypothesis is provided in Secs. 4.2, 5.3,
and 6.

3.3. Cancers

Copper metabolism has been studied in a variety of neoplastic
diseases [38]. It is now known that patients with acute leukemia
have elevated serum or plasma copper concentrations but normal red
blood cell copper levels, although the volume of packed red blood
cells is markedly decreased [39]. The elevation in serum copper
correlated with an increase in bone marrow blast cells. A decline
in symptoms or remission of disease following therapy correlated with
a decrease in serum copper concentration [38,40,41], enabling accurate prognoses based upon serum copper determinations [39].

Chronic leukemia is associated with near-normal plasma copper
concentrations and normal red blood cell concentrations with a
markedly decreased hematocrit. Since remissions do not occur in
cases of chronic lymphocytic and myeloid leukemias as well as mye-
lomas, serum copper levels do not return to normal [38].

Children and adults with active Hodgkin's disease have an
elevated plasma copper concentration and normal red blood cell copper content but, unlike acute and chronic leukemia, these patients
have a normal hematocrit. Copper levels return to normal with remission and increase with relapse, enabling accurate prognosis [42-45].
Since the return to normal copper levels with remission is a constant
feature of this disease, it has been suggested that a normal serum
copper level be included among the criteria for complete remission
[46]. Some patients in remission have been reported to have an elevated serum copper level [47]. It is uncertain what stage of remission these patients had achieved and/or whether they were about to
have a relapse. If a patient had active disease and was entering
into a remission phase, it may require some time before the serum

copper level returns to normal. The increase in serum copper was correlated with an increase in ceruloplasmin concentration [48,49] which was attributed to a lack of catabolism by the liver as opposed to the alternative interpretation of increased ceruloplasmin synthesis.

A good correlation was also found between increased serum copper concentration and disease activity in non-Hodgkin's lymphomas [38]. Patients who responded to therapy had a return to normal serum copper levels, but nonresponders had a persistently elevated serum copper level [50]. Relapse was associated with an elevated copper concentration prior to the onset of relapse symptoms.

Elevation of serum copper levels have also been reported for patients with various carcinomas. The degree of elevation in women with cervical carcinoma increased as the stage of the disease progressed, and those patients who responded favorably to treatment had a nearly normal serum copper level [51]. Patients with bladder carcinoma also have elevated serum copper levels which correlate with the stages of this disease [52]. An elevation in serum copper has also been reported in patients with mammary [38,53], bronchial [54], gastric [55,56], colonic [56], rectal [56], and liver [56] carcinomas as well as osteosarcoma [57]. The degree of elevation in serum copper in osteosarcoma has been correlated with the extent and activity of this disease. The highest copper levels were associated with metastatic disease and the poorest prognosis [58]. Increases in serum copper with liver tumors can be partially accounted for by a failure in liver-mediated copper excretion.

Elevated serum copper levels were also found in dogs with radiation-induced and "spontaneous" osteosarcoma [59]. No clinical signs of metastasis were observed following removal of the tumors by limb amputation, and there was a return to normal or near-normal serum copper levels. Dogs with nonmalignant, nonosteosarcoma lesions were found to have normal serum copper levels.

In each of the above neoplastic diseases, the elevation of serum copper was found to correlate with disease activity, and a return to normal levels with remission. These observations are consistent with the view that the increase in serum copper is a physiological response which may facilitate remission.

The increase in serum or plasma copper concentration found in all cancers studied has special significance since it has recently been recognized that many, if not all, tumor cells have decreased copper-dependent superoxide dismutase (SOD) enzymatic activity [60]. The elevation of blood copper following release from the liver stores may induce SOD or other enzymatic activity in cancer cells and play a role in the facilitation of remission and the subsequent return to normal copper levels. Additional evidence supporting these conclusions is provided in Secs. 4.3 and 6.

3.4. Seizures

Epileptic patients also have elevated blood copper concentrations [61-63]. As shown in Table 1, the brain contains more copper than any other nonstorage tissue in the human body [12], and brain tissues are known to require the following copper-dependent enzymes for normal development and function [4,10,35]:

Cu-dependent enzyme	Role of brain function
Cytochrome-c oxidase	Cellular respiration
Cerebrocuprein (cerebral super-oxide dismutase)	Dismutation of superoxide anion radicals
Tyrosinase	Conversion of tyrosine to Dopa
Dopamine-β-hydroxylase	Conversion of dopamine to norepinephrine
Lysyl oxidase	Conversion of procollagen to tropocollagen and of proelastin to elastin in the vasculature and other connective tissues
Ceramide galactosyl transferase	Myelinogenesis
Soluble pyridoxal-dependent monoamine oxidase	Oxidation of monoamines to aldehydes

In addition, it has been pointed out that copper-dependent processes are required for modulation of prostaglandin syntheses [35], lysosomal membrane stability [35], and the activity of histamine [16], which are also important for normal brain functions.

A symptom of copper deficiency in humans and animals is seizures, which subside with copper supplementation [35,64-69]. Seizures following treatment with tremor-inducing drugs are accompanied by a concomitant reduction in brain copper levels [70-73]. Also, brain norepinephrine and epinephrine concentrations are reduced in association with seizures [10,70,74-79]. This latter observation is particularly relevant since two copper-dependent enzymes are required for the synthesis of norepinephrine and epinephrine.

The hypothesis that seizures result from the loss of copper from some copper-dependent site which may be replaced with the physiological release of liver copper stores is consistent with the above and the observations that (1) inorganic copper injected into the carotid blood supply readily crosses the blood-brain barrier within 15 sec and the amount crossing the barrier is increased by coadministering amino acids [80]; (2) all active antiepileptic drugs [81] are capable of forming copper complexes; and (3) copper complexes have anticonvulsant activity (see Sec. 4.4). Additional support for this hypothesis comes from the observations that postmortem samples of brain tissue from epileptic patients have markedly decreased copper concentrations [64] and that etiologies of epilepsy in children [82] and adults [83] are associated with inflammatory conditions in the central nervous system.

4. PHARMACOLOGIC ACTIVITIES OF COPPER COMPLEXES

4.1. Anti-inflammatory Activity of Copper Complexes

Because of the interest in copper complexes as anti-inflammatory agents and their potential as antiarthritic drugs, a number of reviews have been published recently [35,84,85]. Consequently, this selection will contain only an updated review with the results of the most recent studies.

Prior to 1969, occasional publications reported that sodium-3-(N-allylcuprothiouredo)-1-benzoate [86,87] and cuprous iodide [88] had anti-inflammatory activity and that $Cu(II)(salicylate)_2$ had fever-lowering effects [89,90] in various models of inflammation or fever. In 1969, Bonta [91] and Laroche [92] reported that cupric carbonate $[Cu(OH)_2CuCO_3]$ [91] and cupric complexes of acetic, lauric, oleic, caprylic, butyric, sebasic, lipoic, and cinnamic acids [92] were also effective in animal models of inflammation.

In 1974 [93] and again in 1976 [94,95], it was suggested that copper complexes of clinically used antiarthritic drugs were formed in vivo and that they were responsible for the beneficial antiarthritic effects of these drugs. This suggestion was supported by observations that copper complexes of many non-anti-inflammatory complexing agents, including amino acids, heterocylic carboxylic acids, amines, and other classes of chemical compounds, had anti-inflammatory activity in animal models of inflammation (see Table 2 and Ref. 94).

TABLE 2

Copper Compounds Evaluated for
Anti-inflammatory Activity

Compound	Reference
$Cu(II)Cl_2$	94,103,104,112
$Cu(II)O$	103,104
$Cu(I)Cl$	112
$Cu(I)_2O$	112
$Cu(II)_2(acetate)_4$	94,102-104,112
$Cu(II)(L-tryptophan)_2$	94
$Cu(II)(D-tryptophan)_2$	94
$Cu(II)(anthranilate)_2$	94,112
$Cu(II)(3,5-diisopropylsalicylate)_2$	94

TABLE 2 (Continued)

Compound	Reference
$Cu(II)_2(acetylsalicylate)_4$	94,100-104,112
$Cu(II)(salicylate)_2$	95,101
$Cu(II)_2[2-[3-(trifluoromethyl)phenyl]$ aminonicotinic acid$]_4$	94,102
$Cu(I)_n(D-penicillamine)_n$	94,99,112
$Cu(II)_2(D-penicillamine\ disulfide)_2$	94,99
Cu-D-penicillamine	96,97,106
$Na_5Cu(I)_8Cu(II)_6(D-penicillamine)_{12}Cl$	99
$Cu(II)_2(1-phenyl-5-aminotetrazol)_2(acetate)_4$	94
$Cu(II)_2(1-phenyl-5-aminotetrazol)_4(Cl)_2$	94
$Cu(II)(D-aspartate)$	94
$Cu(II)(L-aspartate)$	94
$Cu(II)(L-lysinate)(Cl)_2$	94
$Cu(II)(pyridine)_2(acetate)_4$	112
$Cu(II)(pyridine)_2(Cl)_2$	94,112
$Cu(II)(morpholine)_2(Cl)_2(HCl)_2$	94
$Cu(II)(1-carboxyisoquinoline)_2$	94
$Cu(II)_2(2-carboxyindole)_3(acetate)$	94
$Cu(II)_3(hydrocortisone-21-phosphate)_2$	94
$Cu(II)_2(hydrocortisone-21-hemisuccinate)_4$	94
$Cu(II)_3(dexamethasone-21-phosphate)_2$	94
$Cu(II)(glycine)_2$	98,99
$Cu(II)-citrate$	98
$Na_2Cu(I)thiomalate$	99
$Cu(I)dithiodiglycol$	99
$Na_2Cu_2(S_2O_2)_2$	99
$Cu(I)(thioacetamide)_4Cl$	99
$Cu(I)(Cu_3CN)_4(CO)_4$	99
$Cu(I)(Cl)(dimethylsulfoxide)$	99
Cu-ascorbate	99
$Cu(II)_2[1-(4-chlorophenyl)-2,5-dimethyl-1H-$ pyrrole-3-acetic acid$]_4$	102

TABLE 2 (Continued)

Compound	Reference
$Cu(II)_2$[2-(3-benzoylphenyl)propionic acid]$_4$	102
$Cu(II)_2$[d-2-(6-methoxy-2-naphthyl)propionic acid]$_4$	102
$Cu(II)_2$[1-(p-chlorobenzoyl)-5-methoxy-2-methyl-3-indolylacetic acid]$_4$	102
$Cu(II)$-2-chloro-4-(o-carboxyphenylamino)pyrimidine	105
$Cu(II)$-N-2,3-dimethylphenylanthranilic acid	105
$Cu(II)$-2,4-di(p-carboxyphenylamino)pyrimidine	105
$Cu(II)$-2,4-di(m-carboxyphenylamino)pyrimidine	105
$Cu(II)$-2,4-di(o-carboxyphenylamino)pyrimidine	105
$Cu(II)$[(-)-2,3,5,6-tetrahydro-6-phenylimidazo(2,1-b)thiazole]$_2$Cl$_2$	106,112
$Cu(II)$-glycyrrhizic acid	107
$Cu(II)$-2[2(6-methoxynaphthalene]propionic acid	108
$Cu(II)$-2(4-isobutylphenyl)propionic acid	109
$Cu(II)$-2(3-benzophenone)propionic acid	110
$Cu(II)$-4-cyclopropylmethyleneoxy-3-chlorophenylacetic acid	111
$Cu(II)$(H$_2$-ethylenediaminetetraacetic acid)	112
$K_2Cu(II)$(ethylenediaminetetraacetic acid)	112
$Cu(II)$(ethylenediamine)$_2$(Cl)$_2$	112
$Cu(II)$diethyldithiocarbamate)$_2$	112
$Cu(II)$[(thiourea)$_3$]$_2$SO$_4$	112
$Cu(I)$(triphenylphosphine)$_3$Cl	112
$Cu(I)$(triphenylphosphine)$_2$(diethyldithiocarbamate)	112
$Cu(II)$(L-histidine)$_2$NO$_3$	112
$Cu(II)$(ethylenediamine)$_2$(NO$_3$)$_2$	112
$Cu(II)$(ethylenediamine)$_2$SO$_4$	112
$Cu(II)$[N-([[(2-phenyl-2-hydroxyethyl)amino]ethyl)-p-toluenesulfonamide]$_2$	112
$Cu(II)$[N-([[(1-phenyl-2-hydroxyethyl)amino]ethyl)-p-toluenesulfonamide]$_2$	112
$Cu(II)$(N-[1-phenyl-2-(2-iminothiazolidin-3-yl)ethyl]-p-toluenesulfonamide)$_2$	112

TABLE 2 (Continued)

Compound	Reference
Cu(II)(pyridine)$_2$(acetylsalicylate)$_2$	112
Cu(I)(triphenylphosphine)(acetylsalicylate)	112
Cu(II)(2-amino-2-thiazoline)$_4$Cl$_2$	112
Cu(II)$_2$(2-[3-(trifluoromethyl)phenyl] aminoanthranilic acid)$_4$	112

In addition, copper complexes of antiarthritic drugs, including salicylic acid, acetylsalicylic acid, 1-phenyl-5-aminotetrazol, 2-[3-(trifluoromethyl)phenyl]aminonicotinic acid, penicillamine, and several corticoids, were found to be more active than the parent drugs. All the copper complexes were found to be more active than either inorganic copper salts or the parent complexing agents regardless of whether or not the parent complexing agent had anti-inflammatory activity.

These observations were confirmed and extended by others [96-99] with the reports that the same and additional copper complexes listed in Table 2 [96-99] had anti-inflammatory activity in an even greater variety of animal models of inflammation. A comparison of copper, gold, and silver thiomalate and thiosulphate complexes in these models of inflammation revealed that the copper complexes were effective while the gold and silver complexes were virtually inactive.

All of the anti-inflammatory activities reported for the compounds listed in Table 2 were obtained following oral or parenteral administration. However, mixtures of copper salts and salicylic acid or ethyl salicylate in aqueous ethanol or glycerol-dimethysulfoxide solutions have recently been shown to have anti-inflammatory activity in a variety of animal models of inflammation following topical application [113-116]. These formulations were developed following the demonstration that copper could be absorbed through the skin [117-119].

The list of compounds presented in Table 2 includes copper complexes of well-known antiarthritic drugs, including salicylic acid, aspirin, niflumic acid, D-penicillamine, hydrocortisone, dexamethasone, dimethylsulfoxide, clopirac, ketoprofen, (+)-naproxen, indomethacin, mefenamic acid, and gold thiomalate. While there has been much discussion concerning the qualitative and quantitative activities of complexes [84,101-104,112,120] it is true that copper complexes of non-anti-inflammatory complexing agents do have anti-inflammatory activity and, where data have been provided, the copper complexes of antiarthritic drugs have been found to be more active or more effective anti-inflammatory agents than the parent drugs. This is in part supported by the observations that many of these and other copper complexes have been studied and found to have antiulcer activity while the parent drugs are ulcerogenic.

4.2. Antiulcer Activity of Copper Complexes

Since it is well known that clinically used antiarthritic drugs cause ulcers and gastrointestinal distress, the observation that copper complexes have antiulcer activity further distinguishes these compounds from their parent ligands as being safer and potentially much more therapeutically useful. The original reports that many copper complexes were effective in preventing ulcers following oral dosing in the Shay and corticoid-induced ulcer models [93-95] have been confirmed in the Shay and other animal models of gastric ulcers. The copper complexes studied to date are listed in Table 3. Based upon these studies, it seems to be true that all copper-containing compounds are orally effective antiulcer agents. Although some are more potent than others, the relative potencies may depend upon the model of ulcer used. There is good agreement in comparisons of results obtained by various investigators.

In addition, some copper complexes of antiarthritic drugs have been studied and found to lack the gastric irritating activity of the parent drugs [100-102]. The lack of gastric irritation, the

TABLE 3

Copper Compounds Evaluated for
Antiulcer Activity

Compound	Reference
$Cu(II)Cl_2$	101,117,126
$Cu(II)_2(acetate)_4$	94,126
$Cu(II)(L-tryptophan)_2$	94,118-119,123,126
$Cu(II)(D-tryptophan)_2$	94,126
$Cu(II)(D,L-tryptophan)_2$	94,126
$Cu(II)(anthranilate)_2$	94
$Cu(II)(3,5-diisopropylsalicylate)_2$	94
$Cu(II)_2(acetylsalicylate)_4$	94,101,102,121-123
$Cu(II)(salicylate)_2$	95,101,121,123,124
$Cu(II)_2(2-[3-(trifluoromethyl)phenyl]$ aminonicotinic acid$]_4$	94,102
$Cu(I)_n(D-penicillamine)_n$	127
$Cu(II)_2(D-penicillamine disulfide)_2$	94,127
$Na_5Cu(I)_8Cu(II)_6(D-penicillamine)_{12}Cl$	127
$Cu(II)_2(1-phenyl-5-aminotetrazol)_2(acetate)_4$	94
$Cu(II)_2(1-phenyl-5-aminotetrazol)_4$	94
$Cu(II)(D-aspartate)$	94
$Cu(II)(L-aspartate)$	94
$Cu(II)(L-lysinate)(Cl)_2$	94
$Cu(II)(epsilon-aminocaproate)$	94
$Cu(II)(pyridine)_2(acetate)_4$	94
$Cu(II)(pyridine)_2(Cl)_2$	94
$Cu(II)(histamine)(Cl)_2(HCl)_2$	94
$Cu(II)_2(nicotinate)_4$	94
$Cu(II)(1-carboxyisoquinoline)_2$	94
$Cu(II)_2(2-carboxyindole)_3(acetate)$	94
$Cu(II)_3(hydrocortisone-21-phosphate)_2$	94
$Cu(II)(2-phenyl-4-carboxyisoquinoline)_2$	94
$Cu(II)(17-hydroxy-3-oxo-17-alpha-pregna-4,$ 6-diene-21-carboxylate$)_2$	94

TABLE 3 (Continued)

Compound	Reference
Cu(II)(4-n-butyl-1,2-diphenyl-3,5-pyrazolidinedione)$_2$	94
Cu(II)$_2$(hydrocortisone-21-hemisuccinate)$_4$	94
Cu(II)$_3$(dexamethasone-21-phosphate)$_2$	94
Cu(II)$_2$[1-(4-chlorophenyl)-2,5-dimethyl-1H-pyrrole-3-acetic acid]$_4$	102
Cu(II)$_2$[1-(p-chlorobenzoyl)-5-methoxy-2-methyl-3-indolylacetic acid]$_4$	94
Cu(II)(L-tryptophan)(L-phenylalaninate)	125
Cu(LL)(L-phenylalanine)$_2$	125
Cu(II)(glycinate)$_2$	128
Cu(II)(L-asparaginate)$_2$	128
Cu(II)(L-alaninate)(L-threoninate)	128
Cu(LL)(L-alaninate)$_2$	128
Cu(II)(L-valininate)$_2$	128
Cu(II)(L-threoninate)$_2$	128
Cu(II)(L-prolinate)$_2$	128
Cu(II)(L-lysinate)$_2$Cl	128
Cu(II)(L-lysinate)(Cl)$_2$	128
Cu(II)(L-histidinate)(L-valinate)	128
Cu(II)(glycinate)(L-histidinate)Cl	128
Cu(II)(L-histidinate)(L-threoninate)	128
Cu(II)(L-glutaminate)$_2$	128
Cu(II)(L-histidinate)$_2$	128
Cu(II)(L-alaninate)(L-histidinate)	128
Cu(II)(L-methioninate)$_2$Cl	128
Cu(II)(L-histidinate)(L-serinate)	128
Cu(II)(L-leucinate)$_2$	128
Cu(II)(L-isoleucinate)$_2$	128
Cu(II)(L-cystinate)	128
Cu(II)(L-seroninate)$_2$	128
Cu(II)(L-alaninatesalicylidene)	128
Cu(II)(anthranilate)(L-phenylalaninate)	128

TABLE 3 (Continued)

Compound	Reference
Cu(II)(anthranilate)(L-tryptophanate)	128
Cu(II)(histaminesalicylidene)(Cl)$_2$	128
NaCu(II)(L-alaninate)(salicylate)	128
Cu(II)(anthranilate)(L-methioninate)	128
Cu(II)(histamine)$_2$(Cl)$_2$	128

presence of antiulcer activity, and the enhanced anti-inflammatory activity of these complexes make this class of potentially useful antiarthritic drugs particularly promising since the arthritic syndrome is likely to include gastric ulcers [36,129].

In general, these complexes have been found to be antisecretory, which may be due to an effect on the modulation of prostaglandin syntheses [102], stimulation of gastric mucous production [121], or modulation of histamine activity [114,125,126,128]. Effective antiulcer activity may be due to an interaction with histamine, which is prosecretory and ulcerogenic, to form copper-histamine which is antisecretory and antiulcer [128]. Copper complexes may also have a role in the promotion of gastric healing [122] or improved connective tissue synthesis and extracellular maturation [124].

In summary, antiarthritic drugs which are known to be ulcerogenic have been converted to antiulcer agents. The antiarthritic drugs which have been studied to date in this regard include salicylic acid, acetylsalicylic acid, phenylbutazone, indomethacin, hydrocortisone, dexamethasone, penicillamine, niflumic acid, ketoprofen, and clopirac.

4.3. Anticancer Activity of Copper Complexes

Sharples [130] was the first to observe that increasing the amount of copper in the diet of rats fed 4-dimethylaminoazobenzene lengthened

hepatic tumor induction time. This observation was confirmed and
extended in subsequent studies [131]. Inorganic copper was also
found to be effective in preventing ethionine-induced liver tumors
in rodents [131] and a variety of other animal carcinomas [132,133].

However, treatment with inorganic copper was not as effective
as therapy with copper complexes. A single 5-mg/kg dose of Cu(II)
(dimethylglyoxime)$_2$ increased the life span of mice with Ehrlich
ascites and sarcoma 180 tumors from 2 to 3 times that of nontreated
controls [132]. Other copper complexes that have been reported to
have similar antitumor activities in rodents are Cu(II)(3,4,7,8-
tetramethyl-1,10-phenanthroline)$_2^{2+}$ [134], Cu(II)(2-keto-3-ethoxy-
butyraldehyde bis-thiosemicarbazone) [135-138], Cu(II)(pyruvaldehyde
bis-thiosemicarbazone) [139], and copper complexes of 2-formylpyri-
dine and 1-formylisoquinoline thiosemicarbazones [140,141], as well
as copper-bleomycin [142].

In addition to demonstrating the antitumor activity of these
copper complexes, the authors cited above have provided a great deal
of information concerning their possible mechanisms of action. All
of these mechanistic studies deal with the inhibition of DNA
synthesis.

However, Oberley and Buettner recently pointed out that all
tumor cell lines have markedly decreased superoxide dismutase (SOD)
enzymatic activity [143]. As a result of the observed lowered SOD
activity in tumor cells and being aware of the SOD-like activity of
our copper complexes (see Sec. 6), we began a study with Oberley and
his colleagues to investigate the antitumor activity of our
complexes.

Oberley et al. [144] observed that copper-dependent SOD obtained from porcine liver was effective in decreasing the growth of
sarcoma 180 cells injected into the hind limb of mice and prolonged
the life of the tumor-bearing mice. They also found that the copper
complex of aspirin, Cu(II)$_2$(acetylsalicylate)$_4$, was more effective
in decreasing tumor growth and prolonging life than the porcine-
derived SOD. In order to increase the lipophilicity of Cu(II)$_2$
(acetylsalicylate)$_4$, which is polymeric, both dimethylsulfoxide and

pyridine monomolecular solvates were prepared. These were found to
be more effective than the polymeric form [144]. Since an increase
in lipid solubility appeared to enhance activity, an ether soluble
complex, Cu(II)(3,5-diisopropylsalicylate)$_2$, was studied and found
to be the most effective antitumor complex we have made to date.

In addition to being effective against sarcoma 180 cells, Cu(II)
(3,5-diisopropylsalicylate)$_2$ also inhibited the growth of Ehrlich
carcinoma cells and markedly increased the life span of the tumor-
bearing mice [145,146]. Increasing the number of treatments further
increased survival time and decreased tumor growth.

Examination of neuroblastoma cells in culture revealed that
"inhibition" of tumor growth was not due to the death of these cells,
but was due to the *differentiation of these neoplastic cell types to
normal cell types*.

Because serum copper concentrations increase in active cancer
and decrease in remission, and copper complexes have anticancer
activity, it seems reasonable to suggest that some copper-dependent
process is required for remission or an anticancer response. Since
the classical approach to cancer chemotherapy has not been successful, there seems to be little risk in pursuing the possibility that
a therapeutically useful copper complex can be found.

4.4. Anticonvulsant Activity of
Copper Complexes

The copper complexes listed in Table 4 have been reported [148,149]
to have anticonvulsant activity in the dose range of 30 to 600 mg/kg
body weight following subcutaneous or intraperitoneal administration
in rodent models of grand mal or petit mal seizures, maximal electro-
shock, or pentylenetetrazol-induced seizures [150-151]. Some of
these compounds had a rapid onset of action, within 30 min following
administration, and the anticonvulsant effect persisted for up to 8
hr. Others had a delayed onset of action, approximately 4 hr, with
a prolonged effect lasting for up to 24 hr. Most of the complexes

TABLE 4

Anticonvulsant Copper Complexes

$Cu(II)_2(acetate)_4$

$Cu(II)_2(adamantylsalicylate)_4$

$Cu(II)_2(acetylsalicylate)_4$

$Cu(II)_2(acetylsalicylate)_4(pyridine)_2$

$Cu(II)_2(acetylsalicylate)_4(dimethylsulfoxide)_2$

$Cu(II)(salicylate)_2$

$Cu(II)(4-aminosalicylate)_2$

$Cu(II)(4-tertiarybutylsalicylate)_2$

$Cu(II)(4-nitrosalicylate)_2$

$Cu(II)(3,5-diisopropylsalicylate)_2$

$Cu(II)(L-threoninate)(L-seriante)$

$Cu(II)(L-threoninate)(L-alaninate)$

$Cu(II)(L-valinate)_2$

$Cu(II)(L-threoninate)_2$

$Cu(II)(L-alaninate)_2$

$Cu(II)(L-phenylalaninate)_2$

$Cu(II)(L-cystinate)_2$

$Cu(II)(L-serinate)_2$

$Cu(II)(L-tryptophanate)_2$

$Cu(LL)(L-glutamate)_2$

$Cu(II)(L-leucinate)_2$

$Cu(II)(L-isoleucinate)_2$

$Cu(II)salicylidene-L-valinate$

$Cu(II)salicylidene-L-histidinate$

$Cu(II)bisacetylacetonethyleneimine$

$Cu(II)bissalicylatoethyleneimine$

Source: Refs. 147-149.

were found to be more effective in the pentylenetetrazol-induced seizure model, but lipophilic complexes were found to be effective in both models of seizure.

At the outset, all compounds were given subcutaneously. Sub-cutaneous administration of 50 to 300 mg/kg of either $Cu(II)Cl_2$ or $Cu(II)_2(acetate)_4(H_2O)_2$ failed to provide any protection against seizures. Recognizing that the ligands used to make these complexes had no anticonvulsant activity (and that some were convulsants) and the lack of anticonvulsant activity following subcutaneous adminis-tration of copper chloride and copper acetate suggested that the ob-served anticonvulsant activity was due to the copper complexes. This led to the suggestion that the active form of the antiepileptic drugs might be their copper complex formed in vivo. In support of this suggestion, we found the copper complex of amobarbital, an anticon-vulsant drug, to be more effective than the parent drug in protecting mice against the maximal electroshock-induced seizures [149]. In addition, it was free of the hypnotic (sleep-inducing) activity of the parent drug if it was absorbed slowly (subcutaneous administra-tion) but was a more potent hypnotic if absorbed rapidly (intraper-itoneal administration). Even $Cu(II)_2(acetate)_4$ had some anticon-vulsant activity following intraperitoneal administration.

Whatever the pathological lesion is that causes epilepsy, it seems clear that studies of the anticonvulsant activities of copper complexes may provide a better approach to therapy of this disease. Also, a better understanding of the altered essential metal metabo-lism in epilepsy may provide a better understanding of the patho-logical lesion.

5. CLINICAL EFFECTIVENESS OF COPPER COMPLEXES IN TREATMENT OF RHEUMATOID AND OTHER DEGENERATIVE DISEASES

5.1. Cupralene and Dicuprene

During the period 1940 to 1950, the copper-containing compounds sodium 3-(N-allylcuprothiouredo)benzoate (Cupralene) and $Cu(II)(4,6-$diethylammonium-8-hydroxyquinoline)$_2$ sulfonate (Dicuprene or Cuprimyl) were reported by Fenz and Forestier (as reviewed in Ref.

152) to be effective in treating a variety of arthritic diseases, including acute and chronic rheumatoid arthritis, chronic polyarticular
synovitis, gonococcal arthritis, chronic polyarticular gout, ankylos-
ing spondylitis, and disseminated spondylitis. Forestier provided
evidence in support of his suggestion that these copper complexes
were superior to gold therapy [152]. Although these results were
confirmed by others [152], the advent of hydrocortisone and the belief that it was going to "cure" arthritis led to the disuse of both
Cupralene and Dicuprene.

5.2. Copper Morrhuate and Alcuprin

Copper morrhuate, a mixture of cod liver oil fatty acid copper complexes, was also evaluated in the treatment of rheumatoid arthritis
during the late 1940s but only 30% of those treated were benefited
[152]. In the early 1950s, the analog of Cupralene, sodium-3-(N-
allycuprothiocarbamide) benzoate (Alcuprin), was equivocally evaluated in the treatment of rheumatoid arthritis and found to be of
little value in treating this disease [152].

5.3. Permalon

Hangarter's preparation of copper salicylate, an aqueous mixture of
12.5 mM of sodium salicylate and 0.04 mM $Cu(II)(Cl)_2$, was more effective than the other clinically used complexes in the treatment of
arthritis and other degenerative diseases. His preparation was effective in treating acute rheumatic fever, acute and chronic rheumatoid arthritis, erythema nodosum, and sciatica either with or without
lower back spinal degenerative disease [152]. A total of 1,140
patients were treated over a 20-year period. Eighty-nine percent
benefited from his therapy, and the average duration of remission of
rheumatoid arthritic disease was 3 years (personal communication).
While Hangarter did not report an antiulcer effect for his

preparation, he did mention a lack of gastrointestinal complaints, which were common when only sodium salicylate was used in therapy. Unfortunately, his approach to therapy was discontinued with his retirement in 1970 and the discontinuation of the manufacture of Permalon for economic reasons. Perhaps another reason for the discontinuance of Hangarter's therapy was a lack of appreciation for why or how copper might be beneficial in the treatment of arthritic diseases.

5.4. Copper Bracelets and Rings

Another likely reason for the lack of interest in the results of Fenz, Forestier, and Hangarter was the folklore use of copper bracelets and rings for the past 5,000 years. Because no beneficial effect could be understood for such devices, they were and still are relegated by some to "quack" medicine. It is unfortunate that the term *quack* has been used when the term *not understood* would have been more appropriate. Since copper bracelets have recently been demonstrated to be effective in relieving arthritic complaints in a single-blind crossover study [117-119], the persistent lay use of copper devices and assertions of beneficial effects seem warranted. Hopefully, another independent study of the copper bracelet will be undertaken to evaluate its effectiveness even if the current state of the art is beyond their use and it may be of no economic value to those funding such a study.

5.5. Ontosein

Most recently, a new anti-inflammatory copper-dependent metalloenzyme, superoxide dismutase (SOD), has been developed for the treatment of arthritic diseases. This drug, Ontosein, has recently been shown to be safe and effective for the treatment of established rheumatoid and osteoarthritis when given intraarticularly into knee

and hip joints in single or multiple doses [153]. Exciting accounts
of the research leading to the discovery of SOD as an anti-inflammatory agent obtained from bovine liver have been published [154-157].
The enzyme is also available for veterinary use as Palosein.

6. BIOCHEMISTRY OF THE ANTI-INFLAMMATORY COPPER COMPLEXES: POSSIBLE MECHANISMS OF ACTION

6.1. Induction of Lysyl Oxidase

It is well known that repair of damaged tissues, resulting in inflammation, requires cross-linking and extracellular maturation of the
connective tissue components collagen and elastin. Since the enzyme
responsible for this, lysyl oxidase, is a copper-dependent enzyme
[158-160], this aspect of wound or tissue repair assumes particular
significance. Harris has demonstrated that lysyl oxidase activity is
induced in copper-deficient chickens with copper(II) sulfate [161].
It may also be possible to induce lysyl oxidase activity with copper
complexes. There is a growing body of evidence that the cofactor
for lysyl oxidase is a pyridoxal-copper complex [162-165]. This
postulated mechanism for copper complex action is consistent with
Townsend's [122] observation that the quantity and quality of the
replaced collagen in $Cu(II)(tryptophan)_2$- and $Cu(II)_2(aspirinate)_4$-
treated rats with surgically induced gastric lesions were superior
and not readily distinguishable from normal in comparison to the
"scarlike" tissue replacement and adhesions found in the nontreated
controls.

6.2. Modulation of Prostaglandin Syntheses

Modulation of prostaglandin syntheses [84] continues to be an increasingly attractive mechanism of action for copper complexes. Inorganic copper salts as well as copper complexes have been shown to
decrease the synthesis of the proinflammatory prostaglandin (PGE_2

vasodilator) and concomitantly increase the synthesis of the anti-
inflammatory prostaglandin (PGF$_{2\alpha}$ vasoconstrictor) [102,166-168].
It is also of interest that rat hearts perfused with pg/ml to μg/ml
concentrations of either PGE$_2$ or PGF$_{2\alpha}$ caused rhythm disturbances
which were completely eliminated when a 2 x 10^{-6} M solution of copper sulfate was added to the perfusate [169].

6.3. Induction of Superoxide Dismutase and
Superoxide Dismutase-Mimetic Activity

Both adult and juvenile rheumatoid arthritis have been associated
with decreased superoxide dismutase (SOD) activity [170,171]. This
is interesting since it is now known that SOD has anti-inflammatory
and antiarthritic activity [153-157] and many of the anti-inflammatory copper complexes studied to date have SOD-like activity [172-
180].

6.4. Stabilization of Lysosomal Membrane

Another possible biochemical mechanism of action for copper complex
action is based upon the report that copper decreased the permeability of human synovial lysosomes by oxidizing membrane thiols to
disulfides (-SH → -S-S-) and, as a result, decreased the release of
free lysosomal enzymes [181]. Membrane stabilization as opposed to
lysosomal enzyme inhibition is also consistent with the observation
that many copper complexes, with the exception of Cu(II)(niflumate)$_2$,
failed to inhibit catespin-D, a lysosomal proteinase [182].

     Stabilization of lysosomes in polymorphonuclear leukocytes
with copper was suggested to account for the reversal of diethyldi-
thiocarbamate-induced toxicity [183]. In the observed biphasic
toxic response, copper protected against the second phase which was
suggested to be due to a reestablishment of the demonstrated loss of
SOD activity. Zinc was more effective in protecting against the
first phase of toxicity, which was suggested to be due to the loss

of enzymatic activity associated with some zinc-dependent enzyme. Since SOD contains both zinc and copper, the first phase may be associated with the removal of zinc from SOD, while the second phase may be associated with the removal of copper from the same enzyme.

6.5. Modulation of Histaminic Activity

Evidence supporting the suggestion that a copper complex of histamine might be the active form of histamine has been published [16,184]. The anaphylactic activity of $Cu(II)(histamine)_2(OH)_2$ was greater than found for the parent ligand alone. The activity of the complex could be decreased by administering zinc salts or a chelating ligand such as salicylate. These results have been supported with the demonstration that $Zn-(lidocaine)(Cl)_2$ and $Zn-(H-lidocaine)_2(Cl)_2$ complexes inhibited compound 48/80-induced histamine "release" from rat mast cells [185]. These complexes also inhibited ionophore A23187- and X537A-induced "releases" of histamine from mast cells. The effects of the corresponding copper complexes were reported to be "weaker" than the zinc complexes, and the $Cu-(H-lidocaine)_2(Cl)_2$ complex "increased" histamine release [186]. It was subsequently reported that the $Cu(II)(histamine)(Cl)(HCl)$ complex and histamine were equiactive in in vitro tests, but the copper complex was always, but not significantly, more active than histamine [187]. Vascular permeability and edema responses were also parallel. In addition, coadministration of ascorbic acid markedly potentiated the effect of the copper complex, while the action of histamine was unchanged. This suggests a role for cuprous copper in modulating histaminic effects since ascorbic acid reduces cupric copper to cuprous.

6.6. Modulation of T-Lymphocyte Responses

Preincubation of human T-lymphocytes with D-penicillamine and only inorganic copper salts decreased their responsiveness (DNA synthesis)

to mitogens but did not alter the capacity of monocytes to act as
accessory cells in response to mitogens [188]. While mixtures of
penicillamine and $Cu(II)Cl_2$, $Cu(I)Cl$, $Cu(II)SO_4$, or $Cu(II)_2(acetate)_4$
were effective in inhibiting this response, mixtures of penicillamine
and $CaCl_2$, $FeSO_4$, $AuCl_3$, or $MgCl_2$ had no inhibitory capacity. Other
mercaptans such as N-acetyl-L-cysteine, cysteamine, D-cysteine, L-
cysteine, 2-mercaptoethanol, or D,L-thiomalic acid were as effective
as D-penicillamine when mixed with inorganic copper salts. It has
also been observed that decreased DNA synthesis in T-lymphocytes
following exposure to diethyldithiocarbamate could be reversed with
inorganic copper [183]. As a result, it has been suggested that the
mechanism of action of D-penicillamine in the treatment of rheumatoid
arthritis involves the formation of D-penicillamine copper complexes
in vivo with the capacity to inhibit T-lymphocyte function [188].
The more recent observation that copper and D-penicillamine also inhibit helper T-cell activity has been used to further support the
efficacy of D-penicillamine in the treatment of arthritis [189].

6.7. Stabilization of Human Gamma Globulin

In another study of the effect of adding copper to D-penicillamine,
it was found that mixture of D-penicillamine disulfide and $Cu(II)SO_4$
prevented heat-induced denaturation of human gamma globulin [190].
Inhibition of synovial fluid gamma globulin denaturation by a copper
D-penicillamine complex was suggested as a plausible mechanism for
the suppression of rheumatoid arthritis.

6.8. Antimicrobial Activity

It has been known for many years that the onset of rheumatoid
disease may follow or occur along with a variety of infections. As
a result, it is of interest that copper complexes have antiviral,
antibacterial, antiprotozoal, antihelmintic, antifungal, and anti-
algal activities [35] as well as antimycoplasmal activity.

Evidence that mycoplasmal infection may play a role in the etiology of rheumatoid diseases has also been provided [191,192]. Since copper complexes of 1,10-phenanthroline, 2,9-dimethyl-1,10- phenanthroline, and 1,3-disubstituted isoquinolines have been shown to be effective antimycoplasmal agents [193,194], this approach to understanding the effectiveness of therapy and the physiological response associated with elevated blood copper levels in arthritic diseases also merits study.

6.9. Modulation of Enzymatic Activities

6.9.1. Preprothrombin Carboxylase

The nicotinamide adenine dinucleotide-dependent (NADH-dependent) generation of superoxide by normal rat liver microsomes is stimulated by the addition of vitamin K, and SOD scavenges the superoxide [195]. Carboxylation of preprothrombin, the precursor of prothrombin, or a pentapeptide composed of amino acids 5 through 9 of prothrombin is inhibited by SOD, but only at high concentrations. This inefficiency of SOD in preventing carboxylation was explained as an inability of SOD to reach the superoxide-generating site [196]. In an attempt to penetrate the barriers to this site, low-molecular-weight SOD-mimetic copper complexes were tested. The copper complexes Cu-D-pencillamine, $Cu(II)(L-tyrosinate)_2$, and $Cu(II)_2$ $(aspirinate)_4$ were found to be more effective than SOD in inhibiting carboxylation. As a result, these low-molecular-weight copper complexes may inhibit the microsomal carboxylating system required for the conversion of preprothrombin to prothrombin and suggest a possible antithrombotic effect for copper complexes with SOD-mimetic activity.

6.9.2. Glutathione-S-Transferase

Changes in liver and intestinal glutathione-S-transferase activities in oleyl alcohol paw edema and adjuvant arthritis inflammations have also been studied [197]. Enzymatic activity was unchanged in acute

inflammation but decreased in chronic adjuvant arthritis. Because
serum copper concentration is elevated in polyarthritis, Cu(II)gly-
cylglycinate was added to tissue homogenates to study the influence
of copper on the amount of transferase activity. Added copper further decreased the level of transferase activity, which may have
been the result of an equilibrium shift from the "free" glutathione-
thiol form to the "bound" disulfide form.

6.9.3. Peptide-Prolyl and Peptide-Lysyl
Hydroxylases, Cytochrome P-450
Mixed-Function Oxidase, NADP-
Cytochrome P-450 Reductase, and
Catechol Dioxygenases

Superoxide dismutase and low-molecular-weight copper complexes have
also been shown to inhibit superoxide anion-dependent biochemical
processes [198-202]. These biochemical processes appear to involve
iron-dependent enzymes which use superoxide to accomplish a variety
of oxidations, including peptide-prolyl and peptide-lysyl hydroxyla-
tion [201], cytochrome P-450 mixed-function oxidation [200] or inhibition of the nicotinamide adenine dinucleotide phosphate-
cytochrome P-450 reductase [201], and catechol dioxygenation [202].
The use of low-molecular-weight copper complexes as probes to study
these enzyme mechanisms is an exciting recent development.

7. CHEMISTRY OF COPPER COMPLEXES

Perrin and his colleagues [203-206] as well as Williams, May, and
their colleagues [1,207-209] developed computer models to study complex formation between low- and high-molecular-weight ligands, such
as amino acids and plasma peptides, and a variety of essential
metals, including copper. With their models, they have been able to
list possible complexes of copper present in blood, the relative
amounts of each, and those responsible for tissue distribution and
excretion of copper.

Computer modeling has also provided experimental evidence in support of the hypothesis that the administration of low-molecular-weight complexes would be beneficial in the treatment of arthritic disease [210]. Studies of the physicochemical binding of anti-arthritic drugs with copper-albumin, a loosely bound form of copper in human blood, demonstrated copper release and the formation of pharmacoactive forms of these drugs [211]. Bonding studies of copper and zinc with albumin demonstrated that both metals were bound and support the generally accepted view that albumin complexes are transportion forms of both metals [212]. Results of studies relevant to cupresis in arthritis patients given D-penicillamine suggested that this drug increased the amount of exchangable copper in plasma by liberating copper from metalloproteins, which was then instrumental in controlling inflammation [213]. Stabilities of copper complexes of salicylic and acetylsalicylic acids were determined to. estimate their concentrations in tissues as they pass through the body. It was concluded that the copper salicylate complex was more absorbable and that the complex dissociated at some point in the process of causing anti-inflammatory and antiulcer effects [214,215]. It was suggested that all effective copper complexes were able to pass through membranes and facilitated tissue distribution of copper.

Additional studies on the chemical influence of D-penicillamine on essential metal metabolism demonstrated that 0.15 g/kg body weight given intramuscularly to rats twice daily for 20 days caused an increased urinary excretion of copper and zinc of 400 and 700%, respectively [216-219]. Statistically, significant reductions of copper, iron, and zinc were also found for all tissues examined: liver, spleen, kidney, heart, lung, brain, and blood (both red blood cells and plasma). In addition, serum ceruloplasmin activity was reduced 69% and cytochrome oxidase activity was reduced 5 to 17% in these tissues. These studies also demonstrated that the weakly bound pool of copper was the source of loss mediated by D-penicillamine, which is consistent with earlier results [220]. It was also noted that D-penicillamine inhibited the oxidation of p-phenylenediamine by ceruloplasmin [217]. Studies with D-cysteine revealed that it was rapidly oxidized by ceruloplasmin, presumably to the disulfide. This observation suggests that some of the loss of tissue metals following treatment with D-penicillamine may be mediated by D-penicillamine disulfide. Alternatively, D-penicillamine disulfide formation may account for a loss of the active form of the drug, the parent mercaptan, and may also partially account for the need for large doses in therapy. If, indeed, D-penicillamine therapy of arthritis causes a net loss of copper and other essential metals, then therapy must be supported with a guaranteed essential metal intake, or, better yet, therapy with the copper complex of D-penicillamine would be best and consistent with the current knowledge.

Complex formation and redox reactions of copper(II) and D-penicillamine would be best and consistent with current knowledge. ligand ratio and concentration of halide ions using electron spin resonance [221]. A pale yellow Cu(I) D-penicillamine polymeric complex is formed when excess D-penicillamine is present. The mixed-valence, red-violet complex is formed when the ratio of D-penicillamine/copper(II) equals 1.45. A very intense blue diamagnetic complex of unknown structure is formed when the D-penicillamine/copper(II) ratio equals 2 and halide ions are present. Complex formation with D-penicillamine and copper(II) was also studied in the presence of albumin, alanine, histidine, and zinc(II). The mixed-valence complex was the major species formed at neutral pH. It was suggested that at concentrations found in blood plasma, it is unlikely that the mixed valence complex or the disulfide complex have any significance in the therapeutic action of D-penicillamine in the treatment of Wilson's disease. Earlier findings [222] that D-penicillamine was unable to mobilize copper bound to albumin were reinforced and it was pointed out that there was some significant non-copper-mediated protein binding, but this could not be used to account for the depletion of copper in Wilson's disease.

Electron spin resonance studies of the structural features of copper carboxylates suggest that the magnetic properties of biologically important molecules which contain copper may be better understood with this spectrophotometric technique [223].

Finally, several excellent reviews of inorganic chemical research dealing with biological systems have been published recently [114,224,225]. With more of this type of chemical information in hand, it may be possible to develop a better pharmacologic approach to the treatment of inflammatory diseases. "It seems possible that the anti-inflammatory action of analgesics, such as aspirin, and the anti-inflammatory action and also the anti-collagen effect of D-penicillamine, may all be linked by their effect on tissue copper, and that collaboration between workers in trace-element metabolism and those in experimental therapeutics would prove fruitful" [226].

ACKNOWLEDGMENTS

I am indebted to the International Copper Research Association, the Arthur Armburst Cancer Research Foundation, and the Arkansas Department of Human Services, Division of Mental Retardation-Developmental Disabilities Services for financial support and to Mr. Gill Gladding and Dr. Harvey Kupferberg of the National Institutes of Neurological and Communicative Disorders and Stroke-Antiepileptic Drug Development Program for the anticonvulsant studies of our complexes.

A special note of appreciation is due to Dr. Larry Oberley and his colleagues Drs. Susan Leuthauser, Garry Buettner, Isabel Bize, and Terry Oberley for a most rewarding collaborative research experience in studying the anticancer activities of copper complexes.

REFERENCES

1. P. M. May, P. W. Linder, and D. R. Williams, *J. Chem. Soc., Dalton Trans.*, 588 (1977).

2. A. Sass-Kortsak, *J. Can. Med. Ass.*, 96, 367 (1967).

3. G. W. Evans, *Physiol. Rev.*, 53, 535 (1973).

4. B. L. O'Dell, in *Trace Elements in Human Health and Disease*, Vol. 1, *Zinc and Copper* (A. S. Prasad and D. Oberleas, eds.), Academic Press, New York, 1976, pp. 391 ff.

5. G. W. Evans, in *Copper in the Environment*, Part 2, *Health Effects* (J. O. Nriagu, ed.), Wiley, New York, 1979, pp. 163 ff.

6. N. Marceau, in *Copper in the Environment*, Part 2, *Health Effects* (J. O. Nriagu, ed.), Wiley, New York, 1979, pp. 177 ff.

7. U. Weser, L. M. Schubotz, and M. Younes, in *Copper in the Environment*, Part 2, *Health Effects* (J. O. Nriagu, ed.), Wiley, New York, 1979, pp. 197 ff.

8. R. Österberg, *Pharm. Ther.*, *9*, 121 (1980).

9. E. Frieden, in *Copper in the Environment*, Part 2, *Health Effects* (J. O. Nriagu, ed.), Wiley, New York, 1979, pp. 241 ff.

10. E. J. Underwood, *Trace Elements in Human and Animal Nutrition*, Academic Press, New York, 1977, pp. 57 ff.

11. G. V. Iyengar, W. E. Kollmer, and H. J. M. Bowen, *The Elemental Composition of Human Tissues and Body Fluids*, Springer, New York, 1978.

12. I. H. Tipton and M. J. Cook, *Health Phys.*, *9*, 105 (1963).

13. P. S. Neumann and A. Sass-Kortsak, *J. Clin. Invest.*, *46*, 646 (1967).

14. R. Osterberg, *Coord. Chem. Rev.*, *12*, 309 (1974).

15. J. R. J. Sorenson, *Progr. Med. Chem.*, *15*, 211 (1978).

16. W. R. Walker, R. Reeves, and D. J. Kay, *Search*, *6*, 134 (1975).

17. J. R. J. Sorenson, *Inorg. Perspect. Biol. Med.*, *2*, 1 (1978).

18. C. A. B. Clemetson, *Lancet*, *2*, 1037 (1968).

19. J. A. Halsted, B. Hackley, and J. C. Smith, *Lancet*, *2*, 278 (1968).

20. H. S. Zacheim and P. Wolf, *J. Invest. Dermatol.*, *58*, 28 (1972).

21. M. D. Barnett and B. Brozovic, *Clin. Chem. Acta*, *58*, 295 (1975).

22. R. S. Pekarek and W. R. Beisel, in *Proceedings of the 9th International Congress on Nutrition, Mexico*, Vol. 2, *1972*, Karger, Basel, 1975, pp. 193 ff.

23. W. R. Beisel, R. S. Pekarek, and R. W. Wannemacher, Jr., in *Trace Elements in Human Health and Disease*, Vol. 1, *Zinc and Copper* (A. S. Prasad and D. Oberleas, eds.), Academic Press, New York, 1976, pp. 87 ff.

24. M. C. Powanda, Trace elements in the pathogenesis and treatment of inflammatory conditions, in *Agents and Action Supplement 8* (K. D. Rainsford, K. Brune, and M. W. Whitehouse, eds.), Birkhäuser, Basel, 1981, pp. 121 ff.

25. R. Hirschelmann, H. Bekemeier, and A. Stelzner, *Z. Rheumatol.*, *36*, 305 (1977).

26. E. J. Gralla and E. H. Wiseman, *Proc. Soc. Exp. Biol. Med.*, *128*, 493 (1968).

27. R. Milanino, S. Mazzoli, E. Passarella, G. Tarter, and G. P. Velo, *Agents Actions*, *8*, 618 (1978).

28. R. Milanino, E. Passarella, and G. P. Velo, *Agents Actions*, *8*, 623 (1978).

29. R. Milanino, E. Passarella, and G. P. Velo, in *Advances in Inflammation Research* (G. Weissmann, ed.), Raven Press, New York, 1979, pp. 281 ff.

30. C. W. Denko, *Agents Actions*, *9*, 333 (1979).

31. I. L. Bonta, in *Inflammation: Mechanisms and Their Impact on Therapy, Agents, and Actions*, Suppl. 3 (I. L. Bonta, J. Thompson, and K. Brune, eds.), Birkhäuser, Basel, 1977, pp. 121 ff.

32. L. L. Wiesel, *Metabolism*, *8*, 256 (1959).

33. P. Koskelo, M. Kekki, E. A. Nikkila, and M. Virkkunen, *Scand. J. Clin. Lab. Invest.*, *19*, 259 (1967).

34. E. W. Rice, *Clin. Chim. Acta*, *5*, 632 (1960).

35. J. R. J. Sorenson, in *Copper in the Environment*, Part 2, *Health Effects* (J. O. Nriagu, ed.), Wiley-Interscience, New York, 1979, pp. 83 ff.

36. W. C. Kuzell, *Ann. Rev. Pharmacol.*, *8*, 368 (1968).

37. J. L. Decker and P. H. Plotz, in *Arthritis and Allied Conditions*, Lea and Febiger, Philadelphia, 1979, pp. 470 ff.

38. G. L. Fisher, in *Proceedings of the 1978 Intra-Science Symposium on Trace Elements in Health and Disease* (N. Kharasch, ed.), Raven Press, New York, 1979, pp. 93 ff.

39. M. Hrgovcic, C. F. Tessmer, B. W. Brown, J. R. Wilbur, D. M. Mumford, F. B. Thomas, C. C. Shullenberger, and G. Taylor, *Prog. Clin. Cancer*, *5*, 121 (1973).

40. H. T. Delves, F. W. Alexander, and H. Lay, *Brit. J. Haematol.*, *24*, 525 (1973).

41. G. Illicin, *Lancet*, *2*, 1036 (1971).

42. C. F. Tessmer, M. Hrgovcic, and J. R. Wilbur, *Cancer*, *31*, 303 (1973).

43. M. Hrgovcic, C. F. Tessmer, F. B. Thomas, L. M. Fuller, J. F. Gamble, and C. C. Shullenberger, *Cancer*, *31*, 1337 (1973).

44. R. L. Warren, A. M. Jelliffe, J. V. Watson, and C. B. Hobbs, *Clin. Radiol.*, *20*, 247 (1969).

45. G. Asbjornsen, *Scand. J. Haematol.*, *22*, 193 (1979).

46. E. Thorling and K. Thorling, *Cancer*, *38*, 225 (1976).

47. J. Wilimas, E. Thompson, and K. L. Smith, *Cancer*, *42*, 1929 (1978).

48. G. L. Fisher and M. Shifrine, *Oncology*, *35*, 22 (1978).

49. P. L. Wolf, G. Ray, and H. Kaplan, *Clin. Biochem.*, *12*, 202 (1979).

50. M. Hrgovcic, C. F. Tessmer, F. B. Thomas, P. S. Ong, J. F. Gamble, and C. C. Shullenberger, *Cancer*, *32*, 1512 (1973).

51. J. A. O'Leary and M. Feldman, *Surg. Forum*, *21*, 411 (1970).

52. L. Albert, E. Hienzsch, J. Arndt, and A. Kriester, *J. Urol.*, *8*, 561 (1972).

53. F. B. DeJorge, J. S. Goes, A. B. Guedes, and A. B. De Ulhoa Cintra, *Clin. Chim. Acta*, *12*, 403 (1965).

54. K. Kolaric, A. Roguljic, and V. Fuss, *Tumori*, *61*, 173 (1975).

55. W. Keiderling and H. Scharpf, *Munsch. Med. Wochenschr.*, *95*, 437 (1954).

56. S. Inutsuka and S. Araki, *Cancer*, *42*, 626 (1978).

57. M. Shifrine and G. L. Fisher, *Cancer*, *38*, 244 (1976).

58. G. L. Fisher, V. S. Byers, M. Shifrine, and A. S. Levin, *Cancer*, *37*, 356 (1976).

59. G. L. Fisher and M. Shifrine, in *Biological Implications of Metals in the Environment* (H. Drucker and R. Wildung, eds.), Research and Development Administration, Technical Information Center of Energy, Oak Ridge, Tenn., 1977, pp. 507 ff.

60. L. W. Oberley and G. R. Buettner, *Cancer Res.*, *39*, 1141 (1979).

61. H. M. Canelas, L. M. Assis, F. B. DeJorge, A. P. M. Tolosa, and A. B. DeUlhoa Cintra, *Acta Neurol. Scand.*, *40*, 97 (1964).

62. C. H. M. Brunia, *Epilepsia*, *7*, 67 (1966).

63. C. H. M. Brunia and B. Buyze, *Epilepsia*, *13*, 621 (1972).

64. V. A. Del'va, *Chem. Abstr.*, *62*, 15160b (1965).

65. A. Ashkenazi, S. Levin, J. Djaldetti, E. Fishel, and D. Benvenisti, *Pediatrics*, *52*, 525 (1973).

66. D. M. Danks, *Inorg. Persp. Biol. Med.*, *1*, 73 (1977).

67. N. T. Griscom, J. N. Craig, and E. B. D. Neuhauser, *Pediatrics*, *48*, 883 (1971).

68. N. Horn, K. Heydorn, E. Damsgaard, I. Tygstrup, and S. Vestermark, *Clin. Genet.*, *14*, 186 (1978).

69. R. A. DiPaolo and P. M. Newberne, in *Trace Substances in Environmental Health*, Vol. 7 (D. D. Hemphill, ed.), University of Missouri Press, Columbia, 1973, pp. 225 ff.

70. D. M. Hunt, *Life Sci.*, *19*, 1913 (1976).

71. S. Hadzovic, R. Kosak, and P. Stern, *J. Neurochem.*, *3*, 1027 (1966).

72. T. R. Price and P. M. Silberfarb, *Amer. J. Psychiat.*, *133*, 235 (1976).

73. P. Stern and S. Kuljak, *Eur. J. Pharmacol.*, *5*, 343 (1969).

74. W. M. Bourn, L. Chin, and A. L.Picchioni, *J. Pharm. Pharmacol.*, *30*, 800 (1978).

75. N. S. Chu, *Epilepsia*, *19*, 603 (1978).

76. S. T. Mason and M. E. Corcoran, *Science*, *203*, 1265 (1979).

77. A. Quattrone, V. Crunelli, and R. Samanin, *Neuropharmacology*, *17*, 643 (1978).

78. T. L. Sourkes, *Pharm. Rev.*, *24*, 349 (1972).

79. P. C. Jobe, A. L. Picchioni, and L. Chin, *J. Pharmacol. Expl. Ther.*, *184*, 1 (1973).

80. J. G. Chutkow, *Proc. Soc. Exp. Biol. Med.*, *158*, 113 (1978).

81. K. W. Leal and A. S. Troupin, *Clin. Chem.*, *23*, 1964 (1977).

82. A. L. Rose and C. T. Lombroso, *Pediatrics*, *45*, 404 (1970).

83. J. M. Oxbury and C. W. M. Whitty, *Brain*, *94*, 733 (1971).

84. I. L. Bonta, M. J. Parnham, J. E. Vincent, and P. C. Bragt, *Progr. Med. Chem.*, *17*, 185 (1980).

85. J. R. J. Sorenson, Trace elements in the pathogenesis and treatment of inflammatory conditions, in *Agents and Actions, Suppl. 8* (K. Brune, K. D. Rainsford, and M. W. Whitehouse, eds.), Birkhäuser, Basel, 1981, pp. 305 ff.

86. W. C. Kuzell, R. W. Schaffarzick, F. A. Mankle, and G. M. Gardner, *Ann. Rheum. Dis.*, *10*, 328 (1951).

87. M. Vykydal, L. Klabusay, and K. Truavsky, *Arzneim. Forsch.*, *6*, 568 (1956).

88. S. S. Adams and R. Cobb, *Progr. Med. Chem.*, *5*, 59 (1967).

89. J. Schubert, in *Metal-Binding in Medicine* (M. J. Seven and L. A. Johnson, eds.), Lippencott, Philadelphia, 1960, pp. 325 ff.

90. J. Schubert, *Sci. Amer.*, *214*, 40 (1966).

91. I. L. Bonta, *Acta Physiol. Pharmacol.*, *Neerl.*, *15*, 188 (1969).

92. M. J. Laroche, *S. A. Bulletin Officiel de la Propriete Industrielle*, No. 3; French patent 6518M (1969).

93. J. R. J. Sorenson, in *Trace Substances in Environmental Health,* Vol. 8 (D. D. Hemphill, ed.), University of Missouri Press, Columbia, 1974, pp. 305 ff.

94. J. R. J. Sorenson, *J. Med. Chem., 19,* 135 (1976).

95. J. R. J. Sorenson, *Inflammation, 1,* 317 (1976).

96. M. W. Whitehouse, L. Field, C. W. Denko, and R. Ryall, *Scand. J. Rheumat., 4, Suppl. 8,* Abstract No. 183 (1975).

97. M. W. Whitehouse, L. Field, C. W. Denko, and R. Ryall, *Agents Actions, 6,* 201 (1976).

98. C. W. Denko and M. W. Whitehouse, *J. Rheum., 3,* 54 (1976).

99. M. W. Whitehouse and W. R. Walker, *Agents Actions, 8,* 85 (1978).

100. D. A. Williams, D. T. Walz, and W. O. Foye, *J. Pharm. Sci., 65,* 126 (1976).

101. K. D. Rainsford and M. W. Whitehouse, *J. Pharm. Pharmacol., 28,* 83 (1976).

102. E. Boyle, P. C. Freeman, A. C. Goudie, F. R. Magan, and M. Thomson, *J. Pharm. Pharmacol., 28,* 865 (1976).

103. A. J. Lewis, *Agents Actions, 8,* 244 (1978).

104. A. J. Lewis, *Proc. Brit. Pharmacol. Soc.,* 413P (1978).

105. A. S. Girgor'eva, E. E. Kriss, S. P. Lazur, S. V. Mikhalovskii, V. A. Portynyagina, N. A. Mokhort, V. K. Kapp, I. S. Barkova, B. A. Kocharovskii, et al., *Khim. Farm. Zh., 12,* 7 (1978); *Chem. Abstr. 89,* 36533c (1978).

106. P. C. Ruenitz, J. R. J. Sorenson, and G. B. West, *Int. Arch. Allergy Appl. Immun., 61,* 114 (1980).

107. F. C. Mondelo, Spanish Patent Application 444,801, January 30, 1976; *Chem. Abstr., 88,* 197661N (1978).

108. D. J. A. Dalmau, Spanish Patent Application 448,954, June 16, 1976.

109. D. J. A. Dalmau, Spanish Patent Application 448,953, June 16, 1976.

110. D. J. A. Dalmau, Spanish Patent Application 450,368, August 2, 1976.

111. G. Pifferi, U.S. Patent 3,976,673, August 24, 1976.

112. D. H. Brown, W. E. Smith, J. W. Teape, and A. J. Lewis, *J. Med. Chem., 23,* 729 (1980).

113. S. J. Beveridge, W. R. Walker, and M. W. Whitehouse, *J. Pharm. Pharmacol., 32,* 425 (1980).

114. W. R. Walker and S. J. Beveridge, *Inorg. Persp. Biol. Med., 2,* 93 (1979).

115. S. J. Beveridge, W. R. Walker, and M. W. Whitehouse, *J. Pharm. Pharmacol.*, *32*, 425 (1980).

116. B. Boettcher, W. R. Walker, and M. W. Whitehouse, Australian Patent Application 77/2,584, November 28, 1977; European Patent Application 2,341, June 13, 1979.

117. W. R. Walker and B. J. Griffin, *Search*, *7*, 100 (1976).

118. W. R. Walker and D. M. Keats, *Agents Actions*, *6*, 454 (1976).

119. W. R. Walker, R. R. Reeves, M. Brosnan, and G. D. Collman, *Bioinorg. Chem.*, *7*, 271 (1977).

120. J. R. J. Sorenson, *J. Pharm. Pharmacol.*, *29*, 450 (1977).

121. K. D. Rainsford and M. W. Whitehouse, *Experientia*, *32*, 1172 (1976).

122. S. F. Townsend and J. R. J. Sorenson, Trace elements in the pathogenesis and treatment of inflammatory conditions, in *Agents and Actions, Suppl. 8* (K. D. Rainsford, K. Brune, and M. W. Whitehouse, eds.), Birkhäuser, Basel, 1981, pp. 389 ff.

123. L. J. Hayden, G. Thomas, and G. B. West, *J. Pharm. Pharmacol.*, *30*, 244 (1978).

124. R. Goburdhun, K. Gulrez, H. Haruna, and G. B. West, *J. Pharmacol. Meth.*, *1*, 109 (1978).

125. F. Marletta, E. Rizzarelli, A. Mangiameli, M. Alberghina, A. Brogna, S. Monaco, S. Sammartano, and A. Blasi, *Boll. Soc. Ital. Biol. Sper.*, *55*, 769 (1979).

126. M. Alberghina, A. Brogna, A. Mangiameli, F. Marletta, E. Rizzarelli, and S. Sammartano, Copper(II) complexes of amino-acids: Gastric acid antisecretory activity in rats, personal communication.

127. J. R. J. Sorenson, K. Ramakrishna, and T. M. Rolniak, Anti-ulcer activity of D-Penicillamine copper complexes, unpublished observation.

128. J. R. J. Sorenson, G. B. West, K. Ramakrishna, and T. M. Rolniak, Antiulcer activity of amino acid copper complexes, unpublished observation.

129. J. L. Decker and P. H. Plotz, in *Arthritis and Allied Conditions* (D. J. McCarthy, ed.), Lea and Febiger, Philadelphia, 1979, pp. 470 ff.

130. G. R. Sharples, *Fed. Proc.*, *5*, 239 (1946).

131. Z. Brada and N. H. Altman, in *Inorganic and Nutritional Aspects of Cancer* (G. N. Schrauzer, ed.), Plenum Press, New York, 1978, pp. 193 ff.

132. I. Donath and G. Putnoky, *Magy. Onkol.*, *13*, 247 (1969); *Cancer Chemother. Abstr.*, *10*(71) (1969).

133. I. V. Savitskii, *Gig. Trans.*, 42 (1970); *Chem. Abstr.*, *74*, 51582b (1970).

134. F. P. Dwyer, E. Mayhew, E. M. F. Roe, and A. Shulman, *Brit. J. Cancer, 19,* 195 (1965).

135. H. G. Petering and G. J. Van Giessen, in *The Biochemistry of Copper* (J. Peisach, P. Aisen, and W. E. Blumberg, eds.), Academic Press, New York, 1966, pp. 197 ff.

136. D. H. Petering and H. G. Petering, in *Antineoplastic and Immunosuppressive Agents,* Vol. 2 (A. C. Sartorelli and D. G. Johns, eds.), Springer, New York, 1975, pp. 841 ff.

137. E. A. Coats, S. R. Milstein, G. Holbein, J. McDonald, R. Reed, and H. G. Petering, *J. Med. Chem., 19,* 131 (1976).

138. D. H. Petering, in *Inorganic and Nutritional Aspects of Cancer* (G. H. Schrauzer, ed.), Plenum Press, New York, 1978, pp. 179 ff.

139. J. G. Cappuccino, S. Banks, G. Brown, M. Searge, and G. S. Tarnowski, *Cancer Res., 27,* 968 (1967).

140. A. Sartorelli and W. A. Creasey, *Ann. Rev. Pharmacol., 9,* 51 (1969).

141. K. C. Agrawal, B. A. Booth, R. L. Michaud, E. C. Moore, and A. Sartorelli, *Biochem. Pharmacol., 23,* 2421 (1971).

142. P. Pietsch, in *Antineoplastic and Immunosuppressive Agents,* Vol. 2 (A. C. Sartorelli and D. G. Johns, eds.), Springer, New York, 1975, pp. 850 ff.

143. L. W. Oberley and G. R. Buettner, *Cancer Res., 39,* 1141 (1979).

144. L. W. Oberley, S. W. C. Leuthauser, G. R. Buettner, J. R. J. Sorenson, T. D. Oberley, and I. B. Bize, *Oxygen Induced Pathology* (A. P. Autor, ed.), Academic Press, New York (in press).

145. S. W. C. Leuthauser, L. W. Oberley, T. D. Oberley, and J. R. J. Sorenson, *J. Nat. Cancer Inst., 66,* 1077 (1981).

146. L. W. Oberley, S. K. Sahu, and J. R. J. Sorenson, Cooper coordination compounds cause morphological differentiation of neuroblastoma cells in vitro, unpublished observation.

147. J. R. J. Sorenson, D. O. Rauls, K. Ramakrishna, R. E. Stull, and A. N. Voldeng, in *Trace Substances in Environmental Health,* Vol. 13 (D. D. Hemphill, ed.), University of Missouri Press, Columbia, 1979, pp. 360 ff.

148. J. R. J. Sorenson, *J. Appl. Nutr., 32,* 4 (1980).

149. J. R. J. Sorenson, R. E. Stull, K. Ramakrishna, B. L. Johnson, E. Riddell, D. F. Ring, T. M. Rolniak, and D. L. Stewart, in *Trace Substances in Environmental Health,* Vol. 14 (D. D. Hemphill, ed.), University of Missouri Press, Columbia, 1980, pp. 252 ff.

150. R. L. Krall, J. K. Penry, H. J. Kupferberg, and E. A. Swinyard, *Epilepsia, 19,* 393 (1978).

151. R. L. Krall, J. K. Penry, B. G. White, H. J. Kupferberg, and E. A. Swinyard, *Epilepsia, 19,* 490 (1978).

152. J. R. J. Sorenson and W. Hangarter, *Inflammation, 2,* 217
      (1977).

153. K. Lund-Olesen and K. B. Menander, *Curr. Ther. Res., 16,* 706
      (1974).

154. W. Huber and M. G. P. Saifer, in *Superoxide and Superoxide
      Dismutases* (A. M. Michelson, J. M. McCord, and I. Fridovich,
      eds.), Academic Press, New York, 1977, pp. 517 ff.

155. W. Huber, K. B. Menander-Huber, M. G. P. Saifer, and P. H.-C.
      Dang, in *Perspectives in Inflammation* (D. A. Willoughby, J.
      P. Giroud, and G. P. Velo, eds.), MPT Press, London, 1977, pp.
      527 ff.

156. K. B. Menander-Huber and W. Huber, in *Superoxide and Superoxide
      Dismutases* (A. M. Michelson, J. M. McCord, and I. Fridovich,
      eds.), Academic Press, New York, 1977, pp. 537 ff.

157. W. Huber and K. B. Menander-Huber, in *Clinics in Rheumatic
      Diseases: Anti-Rheumatic Drugs,* Vol. 2 (E. C. Huskisson, ed.)
      Saunders, London, 1980, pp. 465 ff.

158. W. S. Chou, J. E. Savage, and B. L. O'Dell, *J. Biol. Chem.,
      244,* 5785 (1969).

159. W. H. Carnes, *Fed. Proc., 30,* 995 (1971).

160. B. L. O'Dell, in *Trace Elements in Human Health and Disease,*
      Vol. 1, *Zinc and Copper* (A. S. Prasad and D. Oberleas, eds.),
      Academic Press, New York, 1976, pp. 391 ff.

161. E. D. Harris, *Proc. Nat. Acad. Sci. U.S., 73,* 371 (1976).

162. J. C. Murray and C. I. Levene, *Biochem. J., 167,* 463 (1977).

163. M. S. El-Ezaby and N. El-Shatti, *J. Inorg. Biochem., 10,* 169
      (1979).

164. P. Sharrock, F. Nepveu-Juras, M. Massol, and R. Haran, *Biochem.
      Biophys. Res. Commun., 86,* 428 (1979).

165. M. S. El-Ezaby, H. M. Marafie, and S. Fareed, *J. Inorg. Bio-
      chem., 11,* 317 (1979).

166. R. E. Lee and W. E. M. Lands, *Biochem. Biophys. Acta, 260,* 203
      (1972).

167. I. S. Maddox, *Biochem. Biophys. Acta, 306,* 74 (1973).

168. B. B. Vargaftig, Y. Tranier, and M. Chignard, *Europ. J. Phar-
      macol., 33,* 19 (1975).

169. A. Swift, M. Karmazyn, D. F. Harrobin, M. Monku, R. A. Karmali,
      R. O. Morgan, and A. I. Ally, *Prostaglandins, 15,* 651 (1978).

170. J. M. McCord, *Science, 185,* 529 (1974).

171. M. Rister, K. Bauermeister, U. Gravert, and E. Gladtke, *Lancet,*
      May 20, 1094 (1978).

172. R. Brigelius, R. Spöttl, W. Bors, E. Lengfelder, M. Saran, and U. Weser, *FEBS Lett.*, *47*, 72 (1974).

173. R. Brigelius, H. J. Hartman, W. Bors, M. Saran, E. Lengfelder, and U. Weser, *Hoppe-Seyler's Z. Physiol. Chem.*, *356*, 739 (1975).

174. E. Paschen and U. Weser, *Hoppe-Seyler's Z. Physiol. Chem.*, *356*, 727 (1975).

175. L. R. DeAlvare, K. Goda, and T. Kimura, *Biochem. Biophys. Res. Commun.*, *69*, 687 (1976).

176. U. Weser, C. Richter, A. Wendel, and M. Younes, *Bioinorg. Biophys. Res. Commun.*, *8*, 201 (1978).

177. M. Younes and U. Weser, *Biochem. Biophys. Res. Commun.*, *78*, 1247 (1977).

178. E. Lengfelder and E. F. Elstner, *Hoppe-Seyler's Z. Physiol. Chem.*, *359*, 751 (1978).

179. E. Lengfelder, C. Fuchs, M. Younes, and U. Weser, *Biochem. Biophys. Acta*, *567*, 492 (1979).

180. T. Richardson, *J. Pharm. Pharmacol.*, *28*, 666 (1976).

181. J. Chayen, L. Bitensky, R. G. Butcher, and L. W. Poulter, *Nature*, *222*, 281 (1969).

182. M. H. McAdoo, A. M. Dannenberg, Jr., C. J. Hayes, S. P. James, and J. H. Sanner, *Infect. Immunol.*, *7*, 655 (1973).

183. D. A. Rigas, C. Eqinitis-Rigas, and C. Head, *Biochem. Biophys. Res. Commun.*, *88*, 373 (1979).

184. W. R. Walker and R. Reeves, in *Agents and Actions*, Suppl. 1 (K. D. Rainsford, K. Brune, and M. W. Whitehouse, eds.), Birkhäuser, Basel, 1977, pp. 109 ff.

185. W. Kazimierczak, B. Adams, and C. Maslinski, *Biochem. Pharmacol.*, *27*, 243 (1978).

186. W. Kazimierczak, K. Bankowska, B. Adams, and C. Maslinski, *Agents Actions*, *60*, 59 (1978).

187. A. J. Bennet, T. H. P. Hanahoe, E. J. Lewis, J. R. J. Sorenson, and G. B. West, Histamine Club Meeting, Bergendorf, Sweden, May 16-18, 1979.

188. P. E. Lipsky and M. Ziff, *J. Immunol.*, *120*, 1006 (1978).

189. P. E. Lipsky and M. Ziff, *J. Clin. Invest.*, *65*, 1069 (1980).

190. D. A. Gerber, *Biochem. Pharmacol.*, *27*, 469 (1978).

191. T. M. Brown and H. W. Clark, *INFLO.*, 11 (1978).

192. T. M. Brown and H. W. Clark, *INFLO.*, 12 (1979).

193. B. M. Antic, H. Van der Goot, W. T. Nauta, S. Balt, M. W. G. Debolster, A. H. Stouthamer, H. Verheul, and R. D. Vis, *Eur. J. Med. Chem.*, *12*, 573 (1978).

194. B. M. Antic, H. Van der Goot, W. T. Nauta, P. J. Pijper, S. Balt, M. W. G. Debolster, A. H. Stouthamer, H. Verheul, and R. D. Vis, *Eur. J. Med. Chem.*, *13*, 565 (1978).

195. M. P. Esnouf, M. R. Green, H. A. O. Hill, G. B. Irvine, and S. J. Walter, *Biochem. J.*, *174*, 345 (1978).

196. M. P. Esnouf, M. R. Green, H. A. O. Hill, and S. J. Walter, *FEBS Lett.*, *107*, 146 (1979).

197. E. Fujihira, V. A. Sandeman, and M. W. Whitehouse, *Biochem. Med.*, *22*, 175 (1979).

198. U. Weser and L. M. Schubotz, *Bioinorg. Chem.*, *9*, 505 (1978).

199. R. Myllyla, L. M. Schubotz, U. Weser, and K. I. Kivirikko, *Biochem. Biophys. Res. Commun.*, *89*, 98 (1979).

200. C. Richter, A. Azzi, U. Weser, and A. Wendel, *J. Biol. Chem.*, *252*, 5061 (1977).

201. J. Werringloer, S. Kawano, N. Chacos, and R. W. Estabrook, *J. Biol. Chem.*, 11839 (1979).

202. R. Mayer, J. Wisdom, and L. Que, Jr., *Biochem. Biophys. Res. Commun.*, *92*, 285 (1980).

203. D. D. Perrin, *Nature*, *206*, 170 (1965).

204. D. D. Perrin, *Suomen. Kem.*, *42*, 205 (1969).

205. P. S. Hallman, D. D. Perrin, and A. E. Watt, *Biochem. J.*, *121*, 549 (1971).

206. D. D. Perrin and R. P. Agarwal, in *Metal Ions in Biological Systems*, Vol. 2 (H. Sigel, ed.), Marcel Dekker, New York, 1973, pp. 167 ff.

207. D. R. Williams, *Chem. Rev.*, *72*, 203 (1972).

208. P. M. May, P. W. Linder, and D. R. Williams, *Experientia*, *32*, 1492 (1976).

209. P. M. May and D. R. Williams, *FEBS Lett.*, *78*, 134 (1977).

210. G. E. Jackson, P. M. May, and D. R. Williams, *J. Inorg. Nucl. Chem.*, *40*, 1189 (1978).

211. A. M. Fiabane and D. R. Williams, *J. Inorg. Nucl. Chem.*, *40*, 1195 (1978).

212. A. M. Fiabane, M. L. D. Touche, and D. R. Williams, *J. Inorg. Nucl. Chem.*, *40*, 1201 (1978).

213. M. Micheloni, P. M. May, and D. R. Williams, *J. Inorg. Nucl. Chem.*, *40*, 1209 (1978).

214. G. Arena, G. Kavu, and D. R. Williams, *J. Inorg. Nucl. Chem.*, *40*, 1221 (1978).

215. G. E. Jackson, P. M. May, and D. R. Williams, *J. Inorg. Nucl. Chem.*, *40*, 1227 (1978).

216. V. Albergoni, A. Cassini, N. Favero, and G. P. Rocco, *Biochem. Pharmacol.*, *24*, 1131 (1975).

217. V. Albergoni and A. Cassini, *FEBS Lett.*, *55*, 261 (1975).

218. V. Albergoni, N. Favero, and F. Ghiretti, *Experientia*, *33*, 17 (1977).

219. V. Albergoni, N. Favero, and G. P. Rocco, *Bioinorg. Chem.*, *9*, 431 (1978).

220. G. E. Jackson, P. M. May, and D. R. Williams, *FEBS Lett.*, *90*, 173 (1978).

221. A. Gergely and I. Sóvágó, *Bioinorg. Chem.*, *9*, 47 (1978).

222. S. H. Laurie and D. M. Prime, *J. Inorg. Biochem.*, *11*, 229 (1979).

223. P. Sharrock, M. Dartiquenave, and Y. Dartiquenave, *Bioinorg. Chem.*, *9*, 3 (1978).

224. R. W. Hay and D. R. Williams, *Amino Acids, Peptides Proteins*, *9*, 494 (1978).

225. P. J. Sadler, *Inorg. Perspect. Biol. Med.*, *1*, 233 (1978).

226. M. E. Elms, *Lancet*, November 30, 1239 (1974).

Chapter 5

IRON-CONTAINING DRUGS

David A. Brown
Department of Chemistry
University College
Dublin, Ireland

M. V. Chidambaram
The Florida State University
Tallahassee, Florida

1. INTRODUCTION TO THE CHEMISTRY OF IRON

1.1. Historical Introduction and Importance of Iron(III) as Trace Element

Because of its rich and unique chemical properties and abundance in the earth's crust (among metals it is second only to aluminium), iron has been selected to carry out a wide range of biological functions. In macromolecular environments, iron is responsible for the transport of oxygen (hemoglobin, myoglobin, hemerythrin), the activation of molecular nitrogen and oxygen (nitrogenase and a large variety of oxygenases and oxidases), and electron transport (cytochromes, iron-sulfur proteins). Moreover, because of the propensity of iron to hydrolyze in aqueous solution, special molecules have been designed for both its transport (transferrin) and storage (ferritin).

Only for the (II) and (III) oxidation states of iron are the aqueous ions stable. In acidic solutions ferric and ferrous ions exist as $[Fe(H_2O)_6]^{3+}$ and $[Fe(H_2O)_6]^{2+}$, but as the pH is raised, protons are split off with the formation of hydroxy-iron species [1]. This process, known as hydrolysis, sets in at a neutral pH for ferrous ions and is accompanied by the precipitation of pale green gelatinous $Fe(OH)_2$ which has a solubility product of $10^{-5.1}$ at 25°C and zero ionic strength [2]; this value implies a solubility for Fe^{2+} of about 10^{-1} M at pH 7 and 10^{-3} at pH 8.0. The pK_a of $[Fe(H_2O)_6]^{2+}$ is approximately 7.0 [3]. However iron(III) is much more acidic than iron(II), and the solubility product of $Fe(OH)_3$ is $10^{-38.7}$ at 25°C in 3 M $NaClO_4$ ionic medium [4], which implies a solubility for Fe^{3+} of about 10^{-3} M at pH 2.0 and only 10^{-18} M at pH 7.0. Ferric salts are thus intrinsically unstable in solution unless excess acid

is added; however, with careful techniques, the formation equilibria
of hydroxy complexes which are metastable with respect to precipitation can be studied. The main product [5,6] is a binuclear complex
$[Fe_2(OH)_2]^{4+}$ with smaller amounts of mononuclear species $[Fe(OH)]^{2+}$
and $[Fe(OH)_2]^+$ and a trinuclear complex $[Fe_3(OH)_4]^{5+}$ [7]. Based on
the electronic and magnetic properties of known iron complexes, Gray
has proposed recently [8] that both single and double oxo-bridges as
shown below may occur in hydrolytic iron polymers.

$$Fe \text{———} O \text{———} Fe \qquad\qquad \diagup Fe \diagdown \genfrac{}{}{0pt}{}{O}{O} \diagup Fe \diagdown$$

Iron(III) shows a pronounced tendency toward hydrolytic dimer-
ization even when firmly bound to other ligands [1]. A number of
such dimers have been characterized, some of composition $[(L)Fe(OH)_2]^{2-}$
where L = nitrilotriacetate (NTA^{3-}), N,N'-ethylenediamine tetraacetate
($EDTA^{4-}$), $H(EDTA)^{3-}$, etc. A single-crystal structure determination
has established a nearly linear oxide-bridged structure, i.e.,
LFe—O—FeL for the $H(EDTA)^{3-}$ complex [9]. Infrared and magnetic
data suggest similar structures for the $EDTA^{4-}$, 2,2'-bipyridyl, and
1,10-phenanthroline dimers. Only for the picolinate dimer is there
evidence that the structure involves a double hydroxide bridge [10].

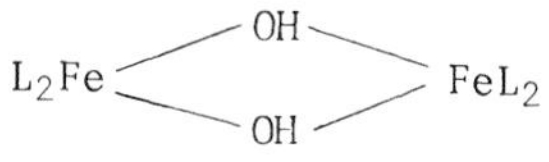

$$L_2Fe \diagup\!\!\!\!\diagdown \genfrac{}{}{0pt}{}{OH}{OH} \diagdown\!\!\!\!\diagup FeL_2$$

The structure of the aquo dimer $[Fe_2(OH)_2]^{4+}$ or $[Fe_2O]^{4+}$ has not
been established, but a double hydroxide bridge is generally assumed
[11]. Dissociation is moderately slow, occurring on a time scale of
seconds, and is acid-dependent [12,13].

1.2. Polymerization of Fe(III) in Solution

Equilibrium measurements on hydroxy-iron(III) complexes are limited
to a narrow composition range. Care must be taken to avoid precipitation; this is most readily achieved by adding base in the form of

HCO_3^-, which forms H_2CO_3 in acidic solutions and is lost as CO_2 leaving behind an equivalent proton deficit [5]. The equilibrium region is limited to solutions containing a hydroxide-to-iron ratio of about 0.5. At higher degrees of hydrolysis, irreversible formation of hydroxy-iron(III) high polymers is encountered [14]. These polymers form clear brown, acid (pH ~2.0) solutions and are stable for long periods of time with respect to precipitation. They are remarkably uniform, with a narrow (~±1%) molecular weight distribution centered around 150,000, corresponding to about 1,200 iron atoms in a particle of average formula $[Fe(OH)_{2.5}]^{0.5+}$, the counter ions being NO_3^-. Electron microscopy reveals that the particles are ~70-Å-diameter spheres. The naturally occurring iron storage molecule ferritin consists of a protein shell surrounding a spherical hydroxy-iron micelle which is also ~70 Å in diameter. Electronic and Mössbauer spectra and magnetic susceptibility measurements show that the ferritin cores have essentially the same internal structure as the hydrolytic polymers [15,16].

When present in equimolar concentration, the citrate ion chelates Fe^{3+} at low pH, with loss of its hydroxide proton. As the solution is neutralized, the characteristic brown color of polymerized iron appears. Alkaline solution (pH ~9.0) contains moderately reactive iron polymers which are again shown by electron microscopy to be ~70-Å-diameter spheres [17]. Precipitation can be avoided, however, if complexing agents are present which bind to the surface of the iron particles and prevent their aggregation. The citrate ions in this case evidently only serve to provide a protective coating, and if sufficient excess citrate (i.e., [Fe]/[citrate] of ~1:20) is present in solution, polymerization is avoided through the formation of a dicitrate chelate [18]. It is worth pointing out that citrate is often used to solubilize iron(III) in biochemical experiments; however, if insufficient citrate is used and the iron in these solutions is polymerized and therefore less reactive, the results of such experiments may be misleading. A good rule of thumb is that if the iron solution is brown, it contains polymers.

2. BIOCHEMISTRY OF IRON

2.1. Iron in Blood, Serum, and Plasma

2.1.1. *General Discussion of the Composition and Function of Blood*

The total blood volume in an adult is ~6 liters (i.e., 7-8% of the
body weight). Approximately 45% of this blood is composed of red
cells (erythrocytes), white cells (leukocytes), and platelets, while
the remaining 55% is the fluid portion termed plasma, of which approximately 90% is water. The remaining 10% of the plasma is composed of proteins (albumin, globulin, and fibrinogen), carbohydrates,
vitamins, hormones, enzymes, lipids, and salts.

Blood may be considered a transport system which, as it cir-
culates through the body, transports oxygen from the lungs to the
tissues, while at the same time products of digestion are absorbed
in the intestine and carried to the various tissues of the body.
Substances produced in various organs are also transferred to other
tissues for use. The cellular elements of the blood are transported
to fight infection or as an aid in blood coagulation in cases of
disease or accident. Waste products from the tissues are picked up
by the blood to be excreted through the skin, kidneys, and lungs.
The separation of blood into serum and plasma is discussed in Sec.
2.1.3.

Blood is found in the interior of the vascular tree, although
fine capillaries provide for increased intimacy between blood and
other tissues that do not normally come into contact with the capillaries. Intestinal tissue fluid is found between the capillaries
and the tissue cells, and this fluid is served by the blood, whereas
the tissues are only served indirectly. The exchange of water and
solutes between the intestinal fluid and the blood is controlled
mainly by a balance of osmotic forces and fluid pressures. As blood
flows through large vessels, the cells are unevenly distributed as
the result of hydrodynamic forces present in a flowing nonturbulent
stream which cause the velocity to be greatest at the axis and least
at the periphery of the vessel. Since the pressure in a moving fluid

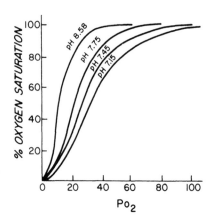

FIG. 1. Oxygen dissociation curve at various pH levels. (From Ref. 19, A. Simmons, *Basic Hematology,* p. 6, 1973. Courtesy of Charles C. Thomas, Publisher, Springfield, Ill.)

is inversely proportional to its velocity, the blood pressure will be lowest at the axis and highest near the vessel wall. Figure 1 shows the solubility of oxygen as a function of its partial pressure, P_{O_2}, and for various blood pH values. It is noticeable that as the pH is lowered (for example, due to increased P_{CO_2}), the P_{O_2} increases. At normal P_{CO_2} values of ~40 mmHg and normal pH values the hemoglobin will be over 90% saturated if P_{O_2} is at least 80 mmHg.

Iron is indispensable in human physiology. The positive charge of iron(III) in solution weakens covalent bonds and makes them more susceptible to cleavage. Biologically active iron-associated compounds may be classified into 10 different functional categories [20], of which the most important are hemoglobin, myoglobin, succinic dehydrogenase, lactic dehydrogenase, xanthine oxidase, transferrin, and ferritin, all of which have well recognized and characteristic functions.

About 65-70% of the total body iron is found in hemoglobin. The turnover rate of hemoglobin is sufficiently rapid to ensure that most of the iron within the body is involved in the hemoglobin cycle. This close relationship between iron and hemoglobin means that disorders of erythropoiesis always affect iron metabolism and, conversely, that disorders of iron metabolism usually affect erythropoiesis.

Mature erythrocytes are composed chiefly of water (~65%) and
hemoglobin (~33% of the wet weight and ~90% of the dry weight); the
remainder consists of numerous enzymes and coenzymes and other proteins, lipids, carbohydrates, amino acids, vitamins, potassium,
sodium, calcium, magnesium, and many trace metals. They also contain other organic compounds such as urea, uric acid, creatine,
creatinine, and various nucleotides [21]. The fully mature, nonnucleated erythrocytes contain no DNA or RNA. Red cell stroma consist primarily of lipids and proteins. The erythrocytes have an
ordered physiological structure and must expend energy to maintain
their integrity and discoidal shape.

Under normal conditions, 1 mm^3 blood contains 4.5-5.5 x 10^6
erythrocytes and in the whole blood ~2.5 x 10^{12} red blood cells [22].
The maintenance of this quantity of erythrocytes on a constant level
is provided for by the production of 200-250 x 10^9 of these cells/
day in the bone marrow. The average life span of an erythrocyte is
~120 days, while 0.83% of the erythrocyte mass is renewed daily.
The normal erythrocyte is a nonnucleated cell having the form of a
biconcave disk with a diameter of 7.9 μm, a thickness of 2 μm, and a
volume of 90 $μm^3$. One erythrocyte contains 30 μγ of hemoglobin.
Respiratory pigment occupies ~90% of the dry weight of the remainder
of the cell. The greatest part of an erythrocyte (~90% of the dry
residue) is made up of hemoglobin. The function of hemoglobin consists of the transfer of gases and composition processes associated
with it.

At present ~60 plasma proteins have been isolated and reasonably characterized [23]. Of these, 10 proteins constitute 84% of
the total protein content of plasma and the remaining 16% is accounted for by the other 50 or more proteins, many of which are
present in extremely low concentration and are referred to as trace
proteins. It is of great interest that of the 60 or more purified
plasma proteins, only prealbumin, albumin, retinol-binding protein,
and four trace proteins are carbohydrate-free; all of the rest contain carbohydrate residues including lipoproteins. In terms of concentration, however, the glycoproteins comprise about 44% of the
proteins because of the high concentration of albumin.

2.1.2. Hemoglobin

Hemoglobin consists of a protein, globin, which constitutes 96% of
the molecular weight, and a prosthetic group, heme (4%). Human
hemoglobin has a molecular weight of 66,700 and contains 0.335%
iron. The molecule is composed of four subunits consisting of heme
and globin and is a spheroid measuring 64 Å by 55 Å by 50 Å [24].
Reduced hemoglobin gives one band at 556 nm in the visible part of
the spectrum, while oxyhemoglobin gives two absorption bands at 542
and 578 nm. Hemoglobin and myoglobin absorb light intensely at
430 and 415 nm, respectively, and these are often referred to as the
Soret bands. Visible absorption is generally due to the heme part
of the molecule, whereas that in the ultraviolet (UV) is due to the
tyrosine and tryptophan residues of the globin. All oxyhemoglobins
(with the exception of hemoglobin M) have identical spectra; however, changes in spectra are produced by substitution of O_2 for CO
or CN, by the oxidation of iron(II) to iron(III), and on denaturation
by acids and alkalies.

Heme is a complex of protoporphyrin IX and iron(II), as shown
in Fig. 2 [22]. The basis of protoporphyrin is the porphyrin ring
constructed from four pyrrole rings connected via methine bridges
with the iron atom being coordinated by the four pyrrole nitrogen

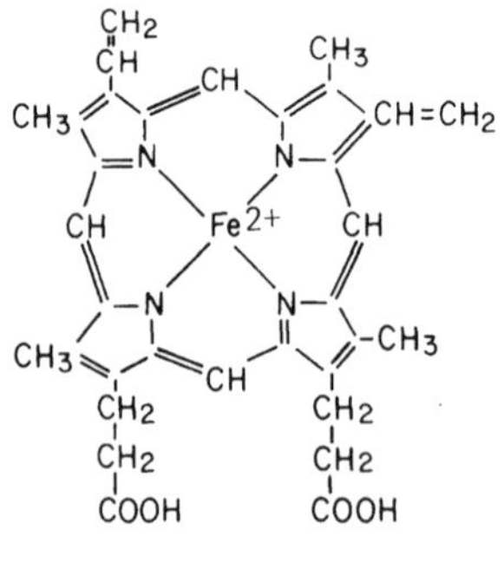

FIG. 2. The heme molecule. (From Ref. 22, J. F. Seitz, *The Biochemistry of the Cells of Blood and Bone Marrow,* p. 10 (1969).
Courtesy of Charles C. Thomas, Publisher, Springfield, Ill.)

atoms. The four-coordinate iron requires two additional ligands to satisfy its normal sixfold coordination number. In hemoproteins, these two bonds are usually formed by histidine groups and may often be replaced by other ligands, e.g., cyanide or carbon monoxide. In hemoglobin, only one of these bonds is formed in this manner by coordination of an imidazole nitrogen of a histidine residue of the polypeptide; this is referred to as the *proximal* side. On the other side (*distal*), the distance is too great for such coordination to occur, and it is this position at which oxygen and other small molecules (CO, CN^-, etc.) can coordinate to complete the normal coordination pattern for iron.

2.1.3. *Serum and Plasma*

One of the very important protective properties of blood is observed if the lining of a blood space is breached by an injury. As the orthopod bleeds the blood begins to ooze away and some of the mobile cells then collect around the gap trying to fill it. Having taken up their places they become immobilized, coalesce, and lose identity as they merge into a plug of jellied protoplasm. In some cases they can also release an agent into the surrounding plasma which causes a precipitation of gluey material which adds further to the bulk of the plug. This is blood clotting (see Fig. 3). The terms *plasma*

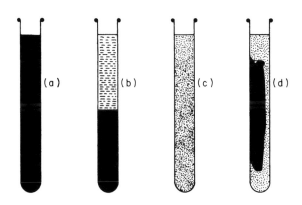

FIG. 3. Blood clotting. (From Ref. 26.)

and *serum* are used when a sample of human blood clots. The clotting protein dissolved in blood (fibrinogen) precipitates as a jelly in which the blood cells become trapped, and as the jelly slowly contracts it holds the cells but exudes a straw-colored juice known as serum. The process is like separating whey from curdled milk, and the term *serum* originally meant whey.

Serum lacks fibrinogen but still contains all the proteins and dissolved substances which have not been involved in the clotting reactions. Much medical testing of blood samples is carried out on serum because it can be obtained so simply free of cells. It also contains all the circulating antibodies, and the laboratory investigation of these is often called *serology*. The word *plasm* as used in *protoplasm* originally meant "formed or molded" but in *plasma* the meaning has changed somewhat. In contrast to serum, plasma contains fibrinogen and many other substances as listed in Sec. 2.1.1, above; consequently, if a sample of plasma is required, an anticoagulent (usually sodium citrate) must be added immediately after the blood is taken to prevent clotting. The pH of the plasma is usually ~7.4, P_{CO_2} is about 40 mmHg, and the pK_a is 5.3 at $37°C$ [25].

During the storage of blood, significant changes take place in the erythrocytes: the concentration of potassium falls, that of sodium rises, the amount of inorganic phosphate (P_i) increases, the content of 2,3-diphosphoglycerate and adenosine 5'-triphosphate (ATP^{4-}) decreases, lactate accumulates, the pH falls from 7.1 to 6.5, glycolytic activity is lowered, lipids are lost, and the concentration of methemoglobin rises. During the third week of storage, in the usual preserving solutions at a temperature higher than $0°C$, the pronounced character and structure of the erythrocytes becomes unstable and they progressively hemolyze. It is possible that the critical role played by the fall in the level of ATP^{4-} is dependent in turn on the changes of activity of a number of enzymatic systems.

From a biochemical point of view, pathological changes in the red part of blood may result from the following causes:

1. Anemias as a result of bleeding.
2. Anemias as a result of a decrease in the hemoglobin content in erthrocytes caused by a disturbance of globin and porphyrin synthesis. Disturbance of iron delivery during disruption or resorption in the gastrointestinal tract and inefficiency of the transport mechanism (transferrin) has also been noted.
3. Anemias conditioned by the increase in destruction of erythrocytes by hemolysis.

A decline of hemoglobin synthesis may also occur despite conditions of an adequate supply of iron and a good resorption in the gastrointestinal tract, as in the case with sideroachrestic anemia. In this case, the defect consists of a disturbance in the mechanism of iron incorporation into heme; thalassemia (major), Colley's anemia is of this type. Megaloblastic anemias are caused by a lack of blood-forming factors such as vitamin B_{12}, folic acid, and mucoprotein of gastric mucin. Pernicious anemia is produced by a blocking in the resorption of vitamin B_{12} as a result of a deficiency of the intrinsic factor, which relates to a defect in the production of HCl in the gastric mucosa. A characteristic manifestation of folic acid deficiency in humans is megaloblastic anemia with extremely peculiar morphological changes in the cells. Quite frequently in anemias produced by hemolysis, the normal life span of erythrocytes is shortened and a wide spectrum of enzymopathies is revealed.

2.2. Fe(III)-Protein Interactions

2.2.1. *Transferrins*

The transferrins comprise a group of homologous proteins present in various fluids of vertebrate animals. All the transferrins form stable complexes with a range of metal ions. The iron complexes are salmon-pink-colored and have been found in the blood serum of all vertebrates studied, in the egg white of all birds, and in the milk of all mammals studied.

The transferrins from blood serum have been called variously the β_1 metal-binding globulin, siderophillin, and transferrin. Holmberg and Laurell [27] proposed that the protein be called transferrin on the basis that its principal function is associated with the transport of iron in serum.

Although transferrin exists in blood sera, milks, and avian egg whites in relatively high concentrations (0.1% to 16% of the solids) only the function of blood serum transferrin has been described satisfactorily. There appears to be little question that its main function is the transport of iron [28-30]. It is also possible that the transferrins may participate in the control of biological processes by regulating the concentration of trace metals, particularly iron. Schade has reviewed previously the earlier studies of the role of serum transferrin in iron transport [30]. Various early investigators observed that blood serum transferrin rapidly binds iron administered through the gastrointestinal tract or by intraveneous injection. There is a rapid turnover of iron in blood serum, and the degree of saturation of the transferrin is related to the amount of iron administered. However, the blood serum transferrin is never fully saturated with iron. Jandl et al. [31] have shown that both ovotransferrin and serum transferrin can transport plasma iron into red cells and that transport is dependent on the concentration of transferrin. Iron taken up by the blood cells cannot be eluted by subsequent incubation with iron-free transferrin solutions. More recently, Morgan and Laurell [29] reported that iron uptake in reticulocytes is independent of the transferrin concentration. It has also been found that the iron complex of serum transferrin has a higher affinity for immature red cells than the iron-free protein. Both bind specifically to immature red cells, thereby permitting the cells to remove iron in the former case. Once the iron is removed, however, the iron-free transferrin can then be replaced by the iron-transferrin complex.

The removal of iron from the complex as outlined above in the case of red blood cells has not been described satisfactorily on a

biochemical basis. Mazur et al. [32] suggested that ATP^{4-} and ascorbic acid are involved in the transfer of iron from transferrin to
ferritin. Although there is a possibility that some type of enzymatic release of the iron may occur, it should be remembered that the
iron transferrin is simply a highly complexed metal chelate and that
physiological conditions such as degree of acidity greatly influence
the dissociation of such a complex. The release of iron-free transferrin in a biological system might therefore be simply a matter of
nonenzymatic dissociation of the complex.

Methods which have been used in the preparation of the various
transferrins include most of those generally applied to the preparation of proteins. The iron complexes have solubilities that differ
from those of the metal-free proteins and are relatively stable compared to many other proteins. The stability of the iron complex is
an advantage because other proteins prevalent as contaminants may be
denatured and insolubilized preferentially. The iron-binding protein of serum transferrin is found in fraction IV-3,4, of human
plasma by low-temperature ethanol fractionation procedure [33,34].
By further sub fractionations, serum transferrin can be concentrated
in Cohn fraction IV-7 [35-37]. Crystallized transferrin is prepared
by crystallization in low dielectric solvents at low temperatures
and controlled ionic strength and pH [38,39]. Large-scale preparation makes use of solvent and salt fractionation and cellulose ion-
exchange chromatography [40] in combination with other procedures
such as electrophoresis [41].

Various physical constants including molecular weights have
been measured by different methods. The S_{20w} values for all transferrins are in the range 5.0 to 6.0, with most of the values clus-
tered around 5.2 to 5.4. Molecular weights range from 86,000 to
93,000. Most transferrins have isoelectric points around pH 6.0.
Metal-free transferrins are relatively stable, especially compared
to the more generally labile proteins such as ovaalbumin [42-44].
Chelation with metal ions apparently causes little change in shape
or size. Human serum transferrin binds Co, Fe, Cu, Zn, and Mn. The

iron complex is red with an absorption maximum at 465 nm with $E_{1cm}^{1\%}$
value of 0.57 at pH 6.5. The importance of CO_2 in the formation of
iron complexes of human serum transferrin was shown at an early date
by Fiala and Burk [45] and confirmed by Warner and Weber [46]. In
the absence of CO_2, the red color of the iron complex forms only
slowly in contrast to its rapid formation in the presence of the
enzyme carbonic anhydrase and free CO_2. The use of $^{14}CO_2$ indicated
that 1 mole of CO_2 was bound per mole of iron bound; however, it is
now realized that carbon dioxide is necessary to form either HCO_3^- or
CO_3^{2-} which bind to the metal.

Davis et al. [47] reported dissociation constants of 10^{-28} and
10^{-30} for the first and second iron atom, respectively, bound in
human serum transferrin in direct contradiction to the results of
Aasa et al. [48], who observed virtually no difference between the
ease of dissociation of the two iron atoms. Electron spin resonance
studies are generally consistent with noninteracting binding sites,
but both Aasa et al. [48] and Windle et al. [41] have shown that
some interaction may occur. Azari and Feeney [42] found that iron
complexes of human serum transferrin are considerably more stable
than their metal free complexes; for example, solutions of the iron
transferrins are more stable to heat, pressure, and the addition of
"denaturing" concentrations of organic solvents or urea than are
solutions of the metal-free proteins. The iron complex of human
serum transferrin is completely resistant to all proteolytic enzymes
tested so far. Azari and Feeney [42] suggested that human serum
transferrin undergoes structural changes on chelation with metal ions
which stabilize the molecule to denaturation and proteolysis. Although it has been suggested that human serum transferrin may consist
of subunits [49], chemical and enzymatic degradation procedures have
not yielded fragments with the original binding properties [42,43],
and there are no sequence data available and very little is known
about the tertiary structure of transferrins. There are no sulfhydryl
groups present, but S—S bonds occur and there is probably one peptide chain and varying amounts of carbohydrate. The lysine content
is in the upper range for proteins varying from 5 to 9 lysine

residues/10,000 molecular weight unit of protein, which implies a
high amino acid content consistent with the reactivity of transferrins to certain reagents. Because of the variable carbohydrate content, transferrins may be classified as glycoproteins; they also
contain sialic acid. The majority of chemical and physical evidence
favors chelation of the iron atoms by 3 tyrosine residues and the
nitrogen atoms of 2 histidine residues with the sixth coordination
position being occupied by either a bicarbonate or carbonate anion.
The search for new variants of human serum transferrins has been
conducted by many workers [50-53], and there are now at least 16
recognized variants of human serum transferrin.

2.2.2. *Ferritin*

The two important iron-containing proteins which together account for
25% of the total iron in humans are ferritin and hemosiderin. In
1894, Schmiedeberg [54] (the well-known German pharmacologist)
described a protein which was isolated from pig liver containing 6%
iron and varying amounts of phosphate. Laufberger [55] described
the isolation of a pure protein rich in iron by heat treatment of
horse spleen followed by repeated precipitation with ammonium sulfate and alcohol. Ferritin is widely distributed throughout nature
and in the various organs of mammals. It is found in relatively
high concentrations in liver, spleen, and bone marrow and in the liver
and spleen of dog, guinea pig, and rat [56]. It represents a depot
in which surplus iron can be stored within the cell in a nontoxic
form and from which it can be mobilized as and when required within
that particular cell or in other cells of the organism. In cells of
the erythroid series (i.e., cells which are precursors of erythrocytes and which synthesize hemoglobin, iron(III) is transported by
transferrin molecules bound to specific receptor sites on the cell
membranes and then transferred into the cells in which, possibly
after reduction to iron(II), it is incorporated into ferritin. At
the end of their life span, erythrocytes are phagocytized by cells
of the spleen, bone marrow, and other tissues. Although the globin

chains are degraded and the porphyrin excreted after conversion to bilirubin in the bile, most of the iron is conserved; it is stored first in ferritin and can be subsequently mobilized from these tissues by transferrin. In liver, muscle, and other tissues, iron is taken up by the cells when transferrin levels are high and deposited first as ferritin and subsequently transferred to hemosiderin. Thus the pool of ferritin iron is used for the synthesis of heme iron proteins. Mobilization of this iron probably involves reduction of iron(III) to iron(II) and its transfer across the cell membrane to transferrin. Ferritin occurs also in the cells of the intestinal mucosa where it has been thought to play a role in regulating the amount of dietary iron absorbed from the gut.

2.2.3. Isolation and Structure of Ferritin

Isolation procedures take advantage of the relative stability of the protein to heat. An aqueous extract is heated to 80°C and after cooling and removal of denatured proteins may be precipitated with ammonium sulfate and purified by repeated crystallization from 5% $CdSO_4$. A number of modifications have been made by various workers [57-62].

Apoferritin can be prepared from ferritin by firstly reducing iron(III) to iron(II) by means of sodium dithionite at pH 4.6-5.0 (e.g., using an acetate buffer) and subsequent separation of the iron(II) by chelation with 2,2'-bipyridyl.

Ferritin consists of a shell of protein subunits surrounding a core of ferric hydroxyphosphate. Electron microscopic data shows clearly that the iron is concentrated in the middle of the apoferritin protein shell [63] as a micelle with a diameter of 120 Å [64-67] which contains ferric iron, predominantly as FeO·OH, and also 1-1.5% of phosphate [57,60]. Mössbauer and infrared spectroscopy and magnetic susceptibility studies of the ferric hydroxy nitrate polymer show that it resembles closely the iron cores of ferritin [68]. A crystal growth model has been proposed by Harrison and co-workers to account for the structure of the micelle [69]. The composition $FeO(OH)_8 \cdot FeO \cdot OPO_3H_2$ has been suggested [70,71].

Harrison and Hoffman [72,73] demonstrated that apoferritin consists of subunits and not one long polypeptide chain. From chemical
studies, a value of 23,000-24,000 for the subunit molecular weight
was calculated [73-75], although a reinvestigation by Crichton and
Bryce using SDS polyacrylamide electrophoresis gave a value of
18,200 ± 1,200 units [76].

2.2.4. *Mobilization of Iron from Ferritin*

The mechanism of the mobilization of iron from ferritin in vivo is
still not clear. The effectiveness of a number of biological reducing agents in mobilizing iron from ferritin was determined by Mazur
et al. [77] and Bielig and Bayer [78]. Mazur and co-workers suggested that an enzyme might be responsible for the reduction and release of ferritin iron, and xanthine oxidase was reported to reduce
ferritin iron under anaerobic conditions in the presence of xanthine
or hypoxanthine [79,80]. However, chelation may also play a part in
the mobilization of iron, possibly in conjunction with reduction,
but the nature of the chelator is not clear. Studies of ferritin
synthesis in duodenal mucosal cells show that oral and parenteral
administration of iron causes a 3-fold and 10-fold increase in the
synthesis of mucosal apoferritin, respectively [81], although this
dosage is in excess of those encountered under physiological conditions. Similar doses were used to show that direct duodenal administration of iron stimulates the synthesis of ferritin by iron-
deficient rats and that this newly synthesized ferritin does not
prevent the uptake of more iron [82].

The effect of iron administration on induction of ferritin in
a number of tissues and in a series of hepatic tumors has been
studied by Linder et al. [83,84].

2.3. Naturally Occurring Nonporphyrin Iron Compounds

This is a short introduction to the relatively low-molecular-weight
nonporphyrin iron compounds which have been isolated from biological

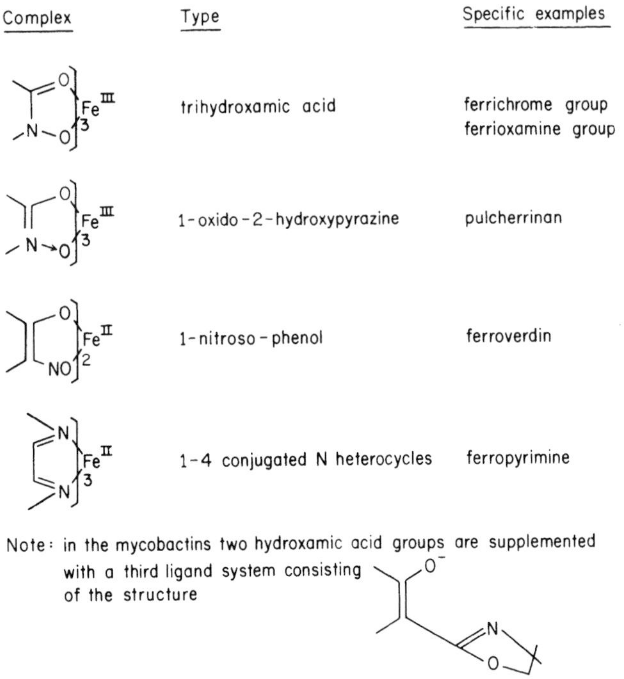

Complex	Type	Specific examples
Fe^{III} (structure)	trihydroxamic acid	ferrichrome group ferrioxamine group
Fe^{III} (structure)	1-oxido-2-hydroxypyrazine	pulcherrinan
Fe^{II} (structure)	1-nitroso-phenol	ferroverdin
Fe^{II} (structure)	1-4 conjugated N heterocycles	ferropyrimine

Note: in the mycobactins two hydroxamic acid groups are supplemented with a third ligand system consisting of the structure

FIG. 4. Ligand types synthesized by microorganisms. (From Ref. 85.)

sources. Apart from the porphyrins, a second major class of iron-complexing ligands, the hydroxamic acids, are found in aerobic microbial cells. The main function of these substances is to carry iron through metabolic channels and insert it into the porphyrins and iron-containing enzymes and proteins. As well as the hydroxamic acids, there are a number of ligand types involved in microbial action as shown in Fig. 4.

2.3.1. *Monohydroxamic Acids*

Hydroxamic acids have attracted the attention of analytical chemists for many years through their ability to form highly colored coordination compounds [86-88]. They are also of considerable biochemical and chemical interest since their basic structure

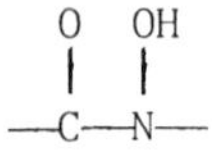

$$\begin{array}{cc} O & OH \\ \| & \| \\ -C & -N- \end{array}$$

occurs in a number of natural products, including antibiotics and
bacterial growth factors. Two excellent reviews on the subject have
appeared. [86,87]. The aceto and benzo derivatives may be regarded
as representative model compounds of the primary monohydroxamic acids,
while the N-methyl derivatives provide examples of simple secondary
monohydroxamic acids.

$$
\begin{array}{cc}
\underset{(Primary)}{\overset{\displaystyle O \quad OH}{\overset{\displaystyle \| \quad |}{CH_3-C-N-H}}} &
\underset{(Secondary)}{\overset{\displaystyle O \quad OH}{\overset{\displaystyle \| \quad |}{C_6H_5-C-N-CH_3}}}
\end{array}
$$

All naturally occurring hydroxamic acids belong to the latter series.
The usual preparation involves the Lossen rearrangement commencing
with the appropriate active acyl precursor [89], e.g., acetohydrox-
amic acid (AHA).

Monohydroxamic acids complex with transition metal ions by
stepwise formation, e.g., with iron(III) the equilibria are as
follows:

$$Fe^{3+} + AHA \rightleftharpoons [Fe(AHA)]^{2+} + H^+$$

$$Fe(AHA)^{2+} + AHA \rightleftharpoons [Fe(AHA)_2]^+ + H^+$$

$$Fe(AHA)_2^+ + AHA \rightleftharpoons [Fe(AHA)_3] + H^+$$

Obviously, the number of hydroxamate anions coordinated to the ferric ion is a function of pH, and this in turn affects both the wave-
length and intensity of absorption of the visible/UV spectrum. At
pH = 1-2, the 1:1 complex is favored and the solution is purple
(λ_{max} = 510 nm; A_{mM} ~ 1.0) [90-92] whereas at neutral pH, the 1:3
complex predominates, giving a yellow-orange color (λ_{max} = 425-440
nm; A_{mM} ~ 3.0). A remarkable property of the hydroxamate ligand is
its pronounced specificity for ferric iron and even the simple monohydroxamic acids, e.g., AHA, have stability constants of ~10^{30}. In
contrast, ferrous iron is bound only relatively weakly, which is
strong evidence against the ferric hydroxamate functioning biologically in the same manner as the hemes, i.e., by alternate oxidation
and reduction. On the other hand, these properties do provide a
simple means of pickup, transfer, and release of iron in living cells.

No metal complex of a monohydroxamic acid has yet been isolated from natural sources, but some interesting and closely related
compounds are worthy of comment--for example, hadacidin:

$$\begin{array}{cc} O & OH \\ \parallel & \mid \\ H\!-\!C\!-\!N\!-\!CH_2\!-\!CO_2H \end{array}$$

which may be isolated and crystallized as its monosodium salt from
the fermentation broth of *Penicillium frequentans*. This compound
shows some interesting biological properties as well as being an
antitumor agent [93].

2.3.2. Trihydroxamic Acids

About a dozen compounds have been isolated from species of fungi and
actinomycetes and contain the same basic structure as in ferrichrome
[94]. In this structure, the hydroxamic acid bond is formed from
acetylated Nδ-hydroxyornithine, whereas in the ferrioxamines it is
derived from a 1-amino-ω-hydroxyaminoalkane. Certain members of
this series, e.g., ferrichrome, are potent growth factors for selected
microbial species, while others such as albomycin and ferrimycin act
as inhibitors [95,96]. Ferrichrome and related compounds can affect
the toxicity of ferric trihydroxamate antibiotics [97,98]. In the
case of acting as growth factors, nutritional requirements may also
be satisfied with higher levels of heme (~1,000 times) or with vast
amounts of inorganic iron, which suggests that ferric trihydroxamates
may act as specific "coenzymes" for transfer of iron in microbial
metabolism. Because of the high specificity of trihydroxamates for
iron, some cases of accidental iron poisoning have been treated successfully by these compounds [95,99] and they have also been used
for the elimination of iron [100]. The structure of ferrichrome-A
obtained by x-ray crystallographic studies [101] is shown in Fig. 5.
The hydroxamate functions are bonded as a cyclic hexapeptide such
that the peptide forms a rough rectangle with a hydrogen bond bridge
(for further details see Ref. 101). Although firmly bound, the iron
in ferrichrome-A undergoes rapid exchange with external iron, illus-
trating the difference between thermodynamic stability and kinetic

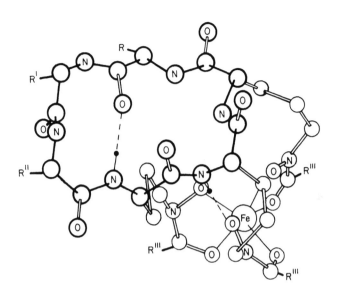

FIG. 5. Structure of ferrichrome. (From Ref. 85.)

reactivity (see Sec. 4). The magnetic moment of ferrichrome-A is
5.73 BM consistent with the presence of Fe(III) [102].

2.3.3. Siderochromes

The term *siderchrome* has been proposed as a general name for all the
red-brown, iron-containing metabolites with a common characteristic
absorption maximum at 540 nm. Similarly *sideramine* and *sideromycin*
may be employed to distinguish further those siderchromes which are
growth factors or antibiotics, respectively. Interestingly, ferrichrome-A exhibits neither of these activities.

2.3.4. Ferrioxamines

Another group of iron-containing trihydroxamate complexes are the
ferrioxamines (see Fig. 6), which are homologs constructed from
chains of amino alkanes, e.g., 1-amino-5-hydroxyaminopentane. They
fall into two groups, linear and cyclic, of which ferrioxamine B and
ferrioxamine E are excellent examples, respectively. The ferrimycins

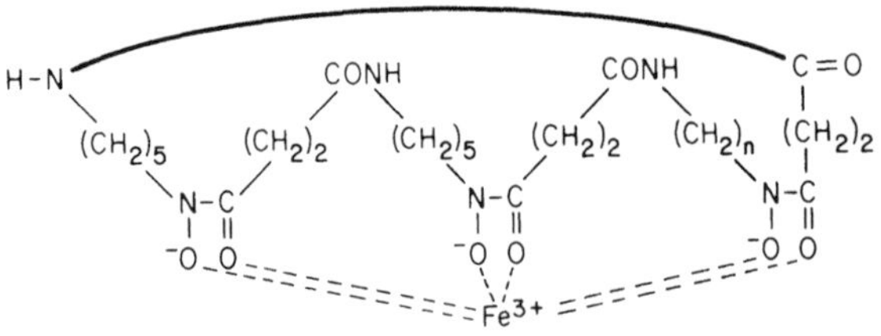

$$H-N \quad\quad\quad CONH \quad\quad CONH \quad\quad C=O$$

$$(CH_2)_5 \quad (CH_2)_2 \quad (CH_2)_5 \quad (CH_2)_2 \quad (CH_2)_n \quad (CH_2)_2$$

$$N-C \quad\quad\quad\quad N-C \quad\quad\quad\quad N-C$$

$$^-O\_O \text{- - - - - - - -} ^-O \quad O \text{- - - - - - - -} ^-O\_O$$

$$Fe^{3+}$$

FIG. 6. General formula of the ferrioxamine group. (From Ref. 85.)

are closely related to the ferrioxamines, but unlike the latter they
display antiobiotic behavior.

Ferrioxamine B is a growth factor for *Microbacterium lacticum*
[103] and reverses inhibition by ferrimycin antibiotics [98].

Further details of the ferrioxamines and ferrichromes can be
found in the excellent review of Neilands [85].

It is obvious from this brief account of naturally occurring
nonheme iron compounds that a large number of "relatively low-
molecular-weight" natural products are capable of binding iron quite
firmly and are associated with it in living cells; indeed, they are
all obtained from microbial sources, suggesting that microorganisms
developed a very extensive coordination chemistry many millions of
years before man! The biological importance of these natural products is still under development.

3. DISEASES BASED ON IRON

3.1. Introduction and Laboratory Diagnosis

The major use of iron-containing drugs is in the treatment of blood
diseases and notably the various anemias. In this section, we give
a brief account of the different clinical conditions couched in lan-
guage which we hope will be appreciated by the nonmedically trained
reader; nevertheless, the use of specialist medical terms is unavoid-
able, so a brief description of each of these is given in the Glos-
sary at the end of this chapter.

Anemia signifies a decreased amount of hemoglobin in the blood and consequently a decreased amount of oxygen reaching the tissues and organs of the body; it is this latter effect which is responsible for many of the symptoms in an anemic person. Anemia has many different causes, and effective treatment cannot be initiated until the exact cause is found.

Iron-deficiency anemia is characterized by a reduction in the erythrocyte count and in the hemoglobin concentration of the peripheral blood, with the latter being proportionately greater.

With mild degrees of anemia (hemoglobin content ~10 to 14 g/100 ml) physiological symptoms are usually not detected but may appear only with heavy exercise reflecting the compensatory overactivity of heart and lungs. They consist of palpitations, dyspnea, and sometimes excessive sweating. With moderate anemia, there is an increase of these symptoms together with fatigue, and finally in severe anemia, tachycardia, wide pulse pressure, hyperpnea, sensitivity to cold, loss of appetite, weakness, and occasional syncope may be seen. In elderly people, local vascular disease may sensitize certain tissues to the effect of anemic hypoxia, and these result in such manifestations as intermittent clandication and angina.

Numerous tests have been devised and in conjunction with the above clinical symptoms are used to differentiate the various types of anemia.

The two most important criteria used to diagnose anemia are (1) the hemoglobin level; and (2) the hemotocrit or packed cell volume (PCV) measurement. This latter quantity is essentially the volume of the blood cells after removal of all plasma, and for standardization purposes a centrifugal force of 2500 g is applied for 30 min to the blood sample when determining the PCV. Closely related to the PCV is the mean corspucular volume (MCV), which is calculated by dividing the volume of packed cells in a given volume of blood by the number of cells in the same volume. The value of the MCV may then be used to classify the anemia as normocytic, microcytic, or macrocytic. Another quantity closely related to the PCV is the

mean corpuscular hemoglobin concentration (MCHC), which is also useful in diagnosis, for example, as an indication of hypochromia.
Most anemias will be normocytic-normochromic, macrocytic-normochro-
mic, or microcytic-hypochromic, depending upon the cause.

Examination of red cell morphology in a stained blood smear is
another basic tool in evaluating anemia and provides an opportunity
to double-check the results obtained from the above red blood cell
indices. For example, the shape of the red cell may be determined
and the presence or absence of sickle cells, spherocytes, target
cells, and inclusion bodies noted. In some anemias, there are special abnormalities present in the white cells or platelets in addition to those in the red cells, and so white cell morphology and
platelet numbers should also be examined. The reticulocyte count
provides an indication of the amount of effective red cell production occurring in the bone marrow but because of inherent errors requires support from other measurements such as the determination of
serum bilirubin, which may indicate increased destruction of red
cells, as in hemolytic anemias. Examination of bone marrow smears
may prove helpful in estimating the relative number of red cells and
their precursors being produced by the marrow. Serum iron and iron-
binding capacity determinations are also helpful as aids to differen-
tiating anemias. The determination of fecal urobilinogen measures
the excretion of the breakdown products of heme, and increased
amounts are generally found in hemolytic anemias and in those
anemias in which ineffective red cell production is present. Three
other tests are worth mentioning although they are used less frequently. First is the plasma iron turnover using ^{59}Fe, in which a
known amount of isotope is injected intraveneously and its rate of
disappearance from the blood is measured by taking blood samples at
regular intervals for 1-2 hr and then determining the amount of
residual radioactivity. In anemias in which the total red cell production is decreased, the ^{59}Fe will remain in the blood longer.
Secondly, red cell utilization of iron is a measure of effective
erythropoiesis, so if the above experiment is monitored for 10-14

days after injection of ^{59}Fe, the latter should normally reappear in the red cells as hemoglobin iron. Thirdly, the life span of the red cell may also be determined by labeling with ^{51}Cr. Blood samples are again taken and the radioactivity measured over a period of 25-35 days, giving a red cell survival curve.

3.2. Classification of Anemias

Since there are numerous causes of anemia, several classifications have been devised by grouping the anemias into different categories.

3.2.1. Iron-Deficiency Anemia

When iron loss exceeds iron intake over a period of time, the iron stores of the body become depleted and there is no longer sufficient iron available for normal hemoglobin production.

The normal adult body contains approximately 4 g of iron, of which about 60% is present in the circulating blood, where 1 ml of red cells contains 1 mg of iron. The remaining body iron is present as stored iron, mainly in the liver and the reticuloendothelial cells of the bone marrow. Each day, 20-25 ml of red blood cells are broken down as a result of normal red cell aging resulting in a daily loss of 1 mg of iron which is excreted through the urine, bile, and other secretions. The remaining 19-24 mg of iron are utilized in the production of more hemoglobin in the formation of red blood cells and so under normal conditions iron-deficiency anemia does not occur. However, if there is an increased need for iron, as in childhood, pregnancy, or excessive blood loss, and iron is thus required by the red cells at a faster rate than is available from the normal dietary intake, the iron stores are utilized for synthesis of hemoglobin iron. When these stores become exhausted iron-deficiency results. It should also be noted that a normal adult can only absorb 5 to 10% of the total daily dietary iron.

In the case of blood loss, the clinical symptoms associated with resulting anemia will depend upon the severity of the bleeding.

In acute blood loss, e.g., when there is a sudden loss of 25-35% of the total blood volume (1-1.5 liters) most healthy patients will show light-headedness and hypotension when they are in an upright position. In chronic blood loss, e.g., when bleeding occurs in small quantities over a period of time, iron-deficiency anemia may develop as a result of the depletion of iron stores. The white count is generally low, as is the reticulocyte count, and polychromatophilia will not be present.

3.2.2. *Hemolytic Anemia (Anemia Due to Destruction of Red Cells)*

This is characterized by a reticulocyte index of 3 or more times normal. In general, a reticulocyte index of less than twice normal indicates a disturbance in erythropoiesis which may be due either to impaired proliferation or to an abnormality in the maturation process, which is associated with excessive cell death prior to the reticulocyte stage resulting also in pigment abnormalities in the anemic patient. Distinction between them is made on the basis of smear tests, various index values, and bilirubin determination.

One group of these anemias is characterized by intravascular hemolysis. A second group is characterized by red cell fragmentation and may have irregularly shaped red cells on smear, the hallmark of these disorders; but by far the largest group is associated with reticuloendothelial destruction of red cells, which may be due to a variety of causes.

Therapy. A decreased oxygen supply due to mild anemia is easily corrected by a slight increase in cardiac output and decreased affinity of Hb for O_2 which is effected by an increase in red cell 2,3-diphosphoglycerate. The deficiency states of particular importance because of their response to specific therapy are iron, folate, and vitamin B_{12} deficiencies. The presence of iron-deficiency anemia usually means blood loss, and it may be more important to identify the cause of bleeding than to treat the anemia. It is important to search for a drug which may have an adverse effect on blood production or destruction.

3.2.3. Hypoproliferative Anemias

This is by far the most frequent type of anemia and arises from one
of the following three main causes: (1) an inadequate iron supply;
(2) decreased stimulation by erythropoeitin; or (3) marrow disease.

The iron supply can be evaluated directly from the plasma iron
and iron-binding capacity, while anemias due to decreased erythro-
poietin stimulation do not show the expected shift cells on smear
tests. Marrow damage is usually associated with abnormalities in
all the formed elements in blood, but marrow examination is often
required to establish the nature of the marrow disfunction, e.g.,
aplaria, neoplastic infiltration, etc.

3.2.4. Maturation Abnormalities

These abnormalities fall into two main categories: (1) macrocytic-
normochromic anemias; and (2) microcytic-hypochromic anemias. The
former are considered to be due to abnormalities in nuclear development and are caused by deficiencies of vitamin B_{12} or folic acid
which may be confirmed by changes in the marrow. Direct evidence of
the two deficiency states may also be obtained by plasma levels.
The latter is caused by iron deficiency or by disorders of globin
metabolism (thalassemia) or protein synthesis (pyridoxine and sideroblastic anemias). The globin and porphyrin abnormalities differ
from iron deficiency in that there is a high plasma iron and accumulation of iron in the developing red cells (sideroblasts).

3.2.5. Megaloblastic Anemias

During the past decade, techniques have been developed that permit
the precise delineation of those vitamins responsible for anemia and
the pathophysiologic factors responsible for the development of these
deficiencies. This condition is due to either vitamin B_{12} or folic
acid deficiency. Vitamin B_{12} is synthesized in nature exclusively
by microorganisms and is present only in wheat and dairy products,
principally in various coenzyme forms. Thus prolonged administration

of such drugs regularly results in the development of megaloblastic anemia; for example, pyrimethamine (Diraprim) acts as a weak folate antagonist and produces a megaloblastic anemia in a high proportion of persons given a dosage of 25 mg/day for 2 months.

The minimum amount of crystalline folic acid available commercially in a capsule is 5 mg. This exceeds the physiological requirement 100-fold. Daily oral administration of this dosage is adequate in all cases of megaloblastic anemia due to folate deficiency. Vitamin B_{12} should be given parenterally. A dosage of 30 µg/month is sufficient to supply the body needs. Initial administration is 1000 µg daily (up to 90% of this dose is lost in the urine) for 5 days, to partially replete hepatic stores in persons with megaloblastic anemia due to deficiency of this vitamin. Maintenance therapy of 100 µg monthly is then adequate.

3.2.6. Sideroblastic Anemias

The sideroblastic or iron-loading anemias are a heterogeneous group of disorders characterized by a disturbance of iron metabolism and frequently result in hypochromic anemia and increased total body iron. The morphological feature defining this group is the presence in the bone marrow of nucleated red cells containing large granules of stainable iron arranged about the cell nucleus--so-called ringed sideroblasts. In normal individuals, the developing bone marrow normoblast takes up iron in excess of its immediate requirements for hemoglobin synthesis. The iron is stored temporarily in the cytoplasm as ferritin or other aggregates which stain with the usual Prussian blue iron stain. Estimation of the proportion of nucleated red cells in normal marrow that contain these granules range up to 90%. The sideroblastic anemias are hereditary disorders and in their primary forms may require phlebotomy therapy to remove excessive iron stores.

3.2.7. *Pregnancy*

Parasitism by the fetus increases folate requirement during pregnancy such that by the third trimester, it approximates to 300 µg/day. Approximately one-third of pregnant women in the United States develop a reduced serum folate concentration during the third trimester and a certain proportion develop a megaloblastic anemia. The administration of certain drugs may also result in megaloblastic anemia due to folate deficiency (see above).

3.3. Recent Advances in Transfusion Therapy

Advances in transfusion therapy have expanded our ability to treat a wide range of conditions. The preparation of blood components, availability of frozen blood, and identification of changes in the erythrocyte during blood storage which may alter its survival and oxygen-carrying capacity are three major areas of development in blood transfusion.

3.3.1. *Red Cells*

Chronic anemia is associated with a compensatory increase in plasma volume, and so infusion of plasma (as a part of whole blood) only contributes to vascular volume overload. Whole blood is necessary only if there is rapid and massive blood loss due to acute hemorrhage. In the case of moderate blood loss (e.g., during surgery) whole blood transfusion is not required, but limited erythrocyte replacement (2-4 units) can be given in the form of packed cells with buffered saline solution. Plasma protein is rapidly repaired in patients who have adequate hepatic function.

3.3.2. *Platelets*

Until multiple packs became available, thrombocytoperic bleeding required transfusions of whole blood. Platelet concentrates prepared from whole blood by low-speed centrifugation may cause

excessive volume expansion and not provide the large number of
platelets required in a short enough time to induce hemostasis.   In
general, when treating deficiencies in soluble clotting factor and
platelet deficiencies, rapid infusion of quantities sufficient to
raise the circulating concentration to at least 25% of normal is a
good goal.   The ability to do this without volume overload is now
possible with the use of concentrates.   Platelet transfusion is not
recommended in the absence of bleeding or even in the presence of
trivial bleeding which can be controlled with local measures.

3.3.3. *Frozen Blood*

Red cell preservation at -80°C permits satisfactory transfusion
after storage for at least 2 and possibly more than 5 years.   This
is in contrast to the present 21-day limit on blood stored at 4°C.
Frozen blood is also useful for transfusion of patients with chronic
renal disease who are candidates for renal transplantation.   Although
the use of additives remains an experimental procedure, studies suggest their potential usefulness in prolonging the shelf life of
stored blood and in improving its immediate oxygen-carrying capacity
in situations in which this could be a crucial factor in the patient's
improvement.

Experience with professional blood donors has shown that most
normal persons can lose about 500 ml of blood every 4-6 weeks without
becoming iron-deficient.

For a selected bibliography see Refs. 19, 25, 158 and 159.

4.   IRON THERAPY

4.1. Dietary Iron and Anemia

There has been much concern about the adequacy of iron in the human
diet, especially since a high incidence of iron-deficiency anemia is
observed throughout the world's population [104-106].   The problem
is especially prevalent in women and children, and revised recommended dietary allowances allowed for the need for increased iron

intake by these groups [107] by increasing dietary iron through oral
enrichment [108]; however, this program has not had any marked effect upon the incidence of anemia, which suggests either insufficient addition or inefficient utilization of iron. Since the dietary intake of iron is often marginal, flour has been supplemented
with iron (13-16.5 mg/lb) in the United States for many years, and
it is proposed to increase this level to 40 mg/lb despite some
criticism. It must be remembered that only 10% of dietary iron is
absorbed, and so the amount of iron ingested must be 10 times the
daily requirements. However, the average American diet provides
only about 6 mg of iron per 1000 kcal [109] and consequently the intake from dietary sources is borderline for teenage girls and women
and may be inadequate for infants and women [110,111]. In the case
of pregnancy, the iron requirements are increased; however, a woman
with sufficient iron stores to provide for her increase in hemoglobin
mass during pregnancy and who breast-feeds her infant for 6 months
will have her iron needs covered by the normal intake of dietary
iron [112].

4.2. Oral Sources of Iron

Oral ferrous sulfate, the least expensive of iron preparations, is
the treatment of choice for iron deficiency, and other sources have
been evaluated relative to $FeSO_4$. Compounds showing a relative biological value of 70 or more are considered good sources (see Fig. 7),
those with a relative biological value between 20 and 70 mediocre
sources, and those with a relative biological value below 20 poor
sources. It is also important to consider the total quantity of iron
furnished. The following are the three categories in which various
iron sources tested have been placed relative to $FeSO_4$.

　　1. *Good sources:* Ferric ammonium citrate, ferric choline
　　　　citrate, ferric chloride, ferric sulfate, ferric ammonium
　　　　sulfate, ferrous fumarate, ferrous gluconate, ferrous

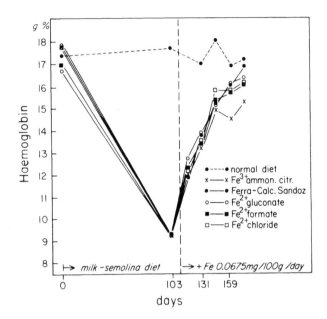

FIG. 7. Experimental anemia in rats--therapeutic experiment: Hemoglobin values in response to a milk and semolina diet and to oral
treatment with various iron preparations. (A biological value of 50
for a sample means that, e.g., 20 mg of iron from the sample gives a
response equal to that obtained from 10 mg of ferrous sulfate in test
animals.) (From Ref. 113.)

sulfate (anhydrous or hydrated), ferrous tartrate, isolated soya bean protein. (For absorption from some of
these sources see Fig. 7.)

2. *Mediocre sources:* Ferric pyrophosphate, reduced iron
(by hydrogen or electrolysis), flour enrichment mixture
(with reduced iron).

3. *Poor sources:* Fe_2O_3, ferric orthophosphate, ferrous
carbonate.

Contrary to many advertisements, gastrointestinal tolerance of
all iron preparations is primarily a function of the total amount of
soluble elemental iron per dose and of physiological factors and is
not normally a function of the form in which iron is administered.
Since iron absorption is critically dependent on the dose administered,

the largest dose without side effects should be given; for most persons, this is 300 mg of hydrated ferrous sulfate or 200 mg of the anhydrous compound thrice daily, but if gastrointestinal symptoms develop, the dose is reduced by one-third or one-half. With such a dosage, the duration of oral therapy for severe iron-deficiency anemia should be approximately 6 months with absorption of ~45 mg of iron daily during the first month when the iron deficiency improves rapidly and subsequent decreasing absorption as the iron stores are slowly used up.

Iron salts have many incompatibilities and so should be prescribed alone, preferably between meals, for maximal absorption, but just after meals if necessary to minimize gastric symptoms. Gastrointestinal absorption of iron is adequate and essentially equal from ferrous sulfate, fumarate, gluconate, succinate, glutamate, and lactate, but lower from ferrous citrate, tartrate, pyrophosphate, cholinisocitrate and carbonate. In general the absorption of iron from ferric salts is poor [114]; reducing agents (e.g., ascorbic acid) and some chelating agents (e.g., succinic acids and sulfur-containing amino acids) may increase absorption, but are probably not worth the extra cost involved.

In the subsequent sections, we list briefly the main properties of known oral sources.

4.2.1. FeSO$_4$ (USP)--Iron Sulfate, FEOSOL

This is the hydrated salt FeSO$_4 \cdot 7H_2O$ containing 20% iron and consisting of pale blue-green crystals or granules. It is soluble in water (1:15 ratio) but oxidizes rapidly in moist air to become coated with brownish yellow basic ferric sulfate, and then it must not be used for medicinal purposes. It is usually dispensed as pills or tablets coated with fructose or lactose as protection against oxidation. The drug is odorless, with a saline astringent taste.

4.2.2. Anhydrous FeSO$_4$ (USP)

This is a grayish white powder containing not less than 80% of the anhydrous salt.

4.2.3. FeSO₄ Syrup (NF)

This contains 40 mg of the salt (8 mg of iron) in each ml. The
average adult dose is 2 teaspoonfuls thrice daily, and for children
who weigh from 15 to 35 kg, 1 teaspoonful twice or thrice daily.

4.2.4. FeSO₄ Oral Solution (USP)

This contains 125 mg (25 mg of iron) in each ml; the average therapeutic dose is 1 ml, three or four times daily. Palatable elixirs
of ferrous sulfate are also available.

4.2.5. Ferrous Fumarate (USP)

This occurs in an anhydrous reddish brown granular powder containing
33% iron and is moderately soluble in water. It is stable and does
not need a protective coating against oxidation. Official ferrous
fumarate tablets usually contain 200 mg of the salt, and the average
adult dosage is 600 to 800 mg daily in divided portions.

4.2.6. Other Ferrous Preparations

Ferrous gluconate (Fergon) contains 12% iron, whereas the lactate
contains 19%; both compounds are tolerated similarly to ferrous sulfate. Ferrous chloride is a light-brown powder readily soluble in
water and containing 44% iron. The formerly recommended dose of 3
or 4 g daily is divided into portions but may cause marked side
effects.

4.2.7. Reduced Iron, Metallic Iron,
        Elemental Iron

For a time these products were the basis of official therapy, but
then interest in them declined, only to revive recently. Indeed,
carbonyl-iron powder, 2-6 μm particle size, has now been suggested
as a new approach to iron deficiency and is possibly the best form
of iron for enrichment of bread and cereal because of apparent
minimal toxicity even when given in massive oral doses (up to 6 g as

a single dose [115]). If further studies support this view, then reduced iron may prove to be an important oral preparation, especially since it would sharply reduce the problem of toxicity to infants who frequently find and ingest an oral preparation meant for an adult.

Sustained-release, delayed-release, and enteric-coated preparations tend to transport iron past the duodenum and proximal jejunum and thus reduce iron absorption.

For a lucid discussion of oral iron therapy see the monograph by Fairbanks et al. [116].

4.2.8. *Toxicity and Side Effects of Oral Preparations [116-119]*

All iron preparations are probably equally toxic per unit of soluble iron. Iron *poisoning* is very rare in adults and unlikely with ingestion of doses of $FeSO_4$ below 50 g. Signs and symptoms may occur within 30 min or may be delayed several hours. They are largely those of gastrointestinal irritation and necrosis with nausea, vomiting, and shock.

Although *side effects* of oral iron therapy occur, there has been much argument concerning their severity and it has even been claimed that they are due to psychological factors [120]. However, it is wise to forewarn patients undergoing therapy that conventional oral doses of iron salts produce constipation in about 10% of cases, diarrhea in about 5%, and nausea and epigastric pain in about 7% (when salts equivalent to 180 mg of ferrous iron per day are given) rising to about 20% (with administration of salts containing 400 mg of ferrous iron per daily dose) [121]. The side effects are primarily a function of the total amount of absorbable iron per dose and may be reduced or eliminated by administration just after meals instead of between meals, since food reduces the absorption of medicinal iron or, more readily, by reducing the dose by one-third to one-half. Finally, large chronic doses of iron may so interfere with the assimilation of phosphorous as to cause severe rickets in infants.

4.3. Parenteral Treatment of Iron Deficiency

Parenteral treatment with intravenous saccharated iron oxide was introduced by Nissam [122] in 1947 for clinical use and was followed shortly afterward by Agner's introduction of intravenous iron dextran [123]. However, saccharated iron oxide lacks stability in the plasma, painful local reactions are developed when injected out of the vein, and a moderate incidence of toxic side effects are noted, depending on the dose injected.

Parenteral iron is indicated only when an iron-deficiency anemia is proved and a trial of oral iron has been found ineffective due to one or more of the following reasons: (1) failure to absorb adequate amounts of oral iron (occasional patients with various malabsorption syndromes); (2) inability to tolerate oral iron (some patients with severe regional enteritis or ulcerative colitis); (3) exhausted iron stores in patients with chronic bleeding, in whom the average daily iron equals or exceeds the absorption of iron from oral ferrous sulfate; (4) refusal or inability to take iron in necessary dosage.

Patients belonging to the first category are rare, the majority being patients with idiopathic steatomahea [124,125] and those who have undergone extensive surgical removal of the small bowel.

In the second group are found some patients with gastrointestinal disorders, especially inflammatory diseases such as ulcerative colitis, regional enteritis, and several other disorders whose symptoms may be aggravated by oral iron salts.

In the third group are included patients losing blood in large amounts, usually from the gastrointestinal tract, for example, as a result of disorders such as hereditary telangiectasia. In these cases, bleeding may be so frequent and severe that one is unable to keep up with the loss by the oral route alone.

In the last category are those people who fail to respond to oral iron therapy for one reason or another, probably due to unsuspected intermittent bleeding in excess of their iron intake.

4.3.1. Preparations Available

There are many parenteral iron preparations available on the market, and the most satisfactory are the following:

1. Saccharated iron oxide
2. Dextriferron
3. Iron dextran (Imferon)
4. Iron-sorbitol-citric acid with dextrin
5. Iron polyisomaltose

The respective intravenous dosages are as follows:

Preparation	Daily dosage, mg
Saccharated iron oxide	
(Ferrivenin)	100-200
(Proferrin)	100
Iron dextrin complex	
(Dextriferron, Astrafer)	100
(Imferon)	100-250

The intramuscular dosages are as follows:

Preparation	Daily dosage, mg
Iron dextran complex (Imferon)	100-500
Iron-sorbitol-citric acid complex (Jactofer)	100
Iron polyisomaltose (Ferrum Hausman Intramuscular)	100

4.3.2. Comparison of Parenteral Preparations

Saccharated iron oxide has been largely superseded by less toxic preparations. When administered intraveneously in a dosage of 100-200 mg, it clears from the plasma in 3-6 hr and is then taken up by the reticuloendothelial system followed by the parenchymal liver cells. The degree of saturation of transferrin will depend on the amount injected. Urinary excretion is negligible.

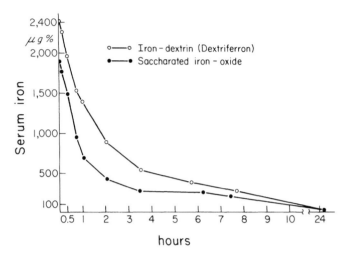

FIG. 8. Serum iron clearance following intravenous injection of 100 mg each of Dextriferron and saccharated iron oxide in a patient with moderately severe iron-deficiency anemia. Interval of 5 days between the two injections. (From Ref. 129.)

Dextriferron is a colloidal solution of Fe(OH)$_3$ complex with partially hydrolyzed dextrin which is a carbohydrate of low molecular weight. It contains 20 mg of iron per ml of isotonic solution and is stable in saline and plasma with a pH of 7.3. It was introduced first in 1948 [123] and has been used by Andersson [126], Lucas and Hagedern [127], and Fielding [128]. Dextriferron is handled in a similar fashion as saccharated iron oxide by the body. The serum iron clearances of these two preparations are compared in Fig. 8.

Iron dextran (Imferon) is a complex of ferric hydroxide and dextran containing 50 mg of iron per ml of solution. Following intramuscular injection, the complex enters the lymph system, first passing through the regional lymph nodes and subsequently slowly absorbed into the bloodstream [130]. Various studies [131,132] using ^{59}Fe show that as much as 25% of the dose may still be present at the site of injection 40-50 days later and is then absorbed very slowly over several months. However, within 2-3 hr the serum iron rises and usually reaches 0.7-1.5 mg%, depending on dosage, and may remain high for several weeks. This preparation has extensive clinical use and is probably the most frequently employed parenteral iron preparation.

Jactofer is a solution of iron-sorbitol-citric acid complex with dextrin as stabilizer. It contains 50 mg Fe per ml at a pH of 7.5 and is stable in serum. Following intramuscular injection, it is promptly absorbed by both blood and lymphatic vessels, raising the serum iron concentration to a maximum of 300-500 µg after 2-3.5 hr [133,134]. It binds rapidly with transferrin to saturate it within 2-3 hr. Using ^{59}Fe it is estimated that 70% of the dose is absorbed in 3 hr, as the utilization of iron for hemoglobin production is very rapid [135,136]. Approximately 30% of the dose is excreted in the urine within 24 hr.

Iron polyisomaltose (Ferrum Hausmann Intramuscular) [137,138] solution contains a complex of ferric hydroxide and polyisomaltose (which forms by the hydrolytic splitting of dextran) and so is closely related to Imferon. It contains 5% iron.

4.3.3. Therapeutic Results

There are many methods for calculating the dosage of iron needed for hemoglobin regeneration. Some prefer the formula which assumes that 25 mg iron is needed to raise the hemoglobin (Hb) level by 1% on the Haldane scale, but the method proposed by Brown and Moore [139] is simple and easy to remember:

(Normal Hb - patient's initial Hb) x 0.255 = grams of iron needed

This method allows for 50% replacement of tissue stores.

In general, it is found that the above three preparations, i.e., Imferon, Jactofer, and Dextriferron, give similar results in treating iron-deficiency anemia. When parenteral iron is given to nonpregnant adults with iron deficiency, the hemoglobin rises from 0.1 to 0.35 g/100 ml per day during the period of maximum hemoglobin production. Using saccharated iron oxide Coleman et al. [140] found a value of 0.25 g Hb per 100 ml per day when the initial hemoglobin level was below 7 g, and in one investigation using blood volume

studies it was reported that over 20% of the patients showed values
of 0.35 to 0.40 g Hb per 100 ml per day [141]. The higher values
for hemoglobin regeneration occur when larger doses of iron (200-500
mg) are administered; the hemoglobin response in children appears to
be somewhat greater.

It appears that comparable hemoglobin regeneration rates can
be obtained with parenteral and oral iron therapy.

4.3.4. Toxicity of Parenteral Iron Preparations

The toxicity of the various preparations has caused difficulties,
and the main toxic reactions fall into (1) the local type; and (2)
systemic types. These are listed below:

> *Side Effects of Parenteral Iron*

1. *Local pain at the site of injection*
   Skin decoloration
   Tender inguinal lymphadenopathy, abdominal pain in
   the lower quadrant
   Sarcoma

2. *Systemic reactions*
   Fever chills
   Headache, nausea, vomiting
   Flushing of the face, hypotension, dizziness

The toxicity of saccharated iron oxide is fully documented
[139], while that of Imferon has been estimated at around 0.5% by
Scott and Govan [142], who caution against its use in patients with
asthma or skin rashes. However, in 1960 Imferon was temporarily
withdrawn from the market following the reports by Richmond [143,144]
and Haddow and Horning [145] of the induction of sarcomas in rats and
mice at the side of injection after repeated massive doses of the
drug. Nevertheless, it is now believed that the danger of producing
malignancy in humans is very remote, and Imferon accordingly reap-
peared on the market in the United States in 1962.

Dextriferran toxicity has been reported to be mild and infre-
quent. Leakage of the preparation from the vein results in minor

inflammation which usually subsides in 24-48 hr. Flushing, nausea,
headaches, and abdominal pain may occur. Delayed side effects include a feeling of stiffness in the extremities or the face, chills,
and fever. Recently severe reactions were documented in two patients
receiving the preparations while on oral iron therapy [146].

Initial studies with the iron-sorbitol-citric acid complex indicated only mild toxicity consisting mainly of local pain at the
side of injection. It may also interfere with the sense of taste
starting about 1 hr after injection and lasting for several hours.
It is probably best not to give iron-sorbitol in association with
oral iron.

Because of the mild to severe toxic reactions which may develop
with all the parenteral iron preparations presently available, they
should only be used with certain clear indications in mind. In the
absence of a clear need, it appears that oral treatment can be
equally effective.

4.4. Therapy for Cooley's Anemia (Iron Overload)

As mentioned in Sec. 3.1, microcytic-hypochromic anemias lead to the
problem of iron overload (e.g., Cooley's anemia) which results in
the saturation of transferrin in the system and toxicity since the
excess iron leaves the bloodstream and accumulates in the tissues.
Successful treatment of this condition involves the use of chelating
agents and will probably have to be of lifetime application. This
problem has been reviewed by Waxman and Brown [147] who suggested
that the ideal drug should be:

1. Inexpensive
2. Orally administrable
3. Nontoxic
4. Resistant to degradation prior to efficient absorption
    via the gastrointestinal tract
5. Able to disperse through the bloodstream and bind iron
    in competition with transferrin

The most effective drugs presently available for long-term treatment of iron overload are (1) desferrioxamine-B (DFB), usually administered as the methylsulfonate salt (Desferal, Ciba-Geigy); and (2) diethylenetriamine-pentaacetic acid (DTPA). There is much current research in this area involving the design of suitable chelating agents, the subject of a recent symposium [148]. It has also been noted recently that 2,3-dihydroxybenzoic acid is an effective agent in the promotion of iron excretion in the hypertransfused rat [149].

4.5. Criteria for Biological Activity in Iron Chelates

In principle, the design of an iron-containing drug for treatment of the various diseases discussed in Sec. 3 depends on the development of suitable chemical criteria from in vitro experiments. However, although we are still far from this ideal position, much interest has grown in recent years regarding the design of iron chelates for the oral treatment of iron deficiency [150,151] and the closely related problem of the design of chelates for treatment of iron overload [147,148].

The successful prediction of biological activity for a given chelate is a difficult task and depends on an understanding of its function in biological systems and our ability to relate these functions to simple physicochemical properties of the chelate measured in vitro. The classic work of Neilands and of Saltman provided early examples of this type of approach in which they stressed that the thermodynamic stability of metal complexes depends not only on the properties of the ligands but also on those of the metal and the type of metal-ligand bonds formed; moreover, high thermodynamic stability does not necessarily indicate kinetic inertness [152]. In biological media, it is very important to establish the kinetic stability of the iron complex, since it is this property which will determine the rate of ligand substitution reactions between the complex and the many active chelating ligands present in mammalian systems. This point is well illustrated by the work of Bates and Wernicke [153] who studied the kinetics and mechanism of Fe(III) exchange between iron chelates and transferrin where, despite relatively small differences in thermodynamic stability constants, very large differences in rates of reaction were observed.

In the past few years, we have undertaken a joint research program with agricultural scientists aimed at developing better oral iron sources for various mammals. We have suggested a set of physicochemical criteria for biological activity and tested these for a number of chelates in terms of the ability of a given chelate to regenerate hemoglobin levels in anemic rats [150]. The following criteria for biological activity are suggested:

1. *Thermodynamic:*

    The chelate must be stable and monomeric at biological pH values in order that it may cross various cell membranes.

2. *Kinetic:*

    a. It should be able to exchange iron rapidly with apotransferrin, since transferrin is responsible for transport in mammalian systems.

    b. The chelating ligand should be able to extract iron fairly rapidly from ferritin, the iron storage protein.

These criteria have been applied recently to a series of iron(III) chelates of monohydroxamic acids [150] and amino hydroxamic acids [151]. The application of the first kinetic criterion based on transfer of ferric ions between the chelate and apotransferrin is given in Table 1. It is immediately obvious that both ferric acetohydroxamate (AHA) and ferric glycinehydroxamate (GHA) exchange iron very rapidly, and so biological activity is strongly indicated. The prediction for Fe(AHA)$_3$ is fully indicated by animal studies which showed this chelate to be slightly superior to ferric citrate in the regeneration of hemoglobin levels in anemic rats [150]. This program of research is continuing.

TABLE 1

Rate of Fe^{3+} Exchange with Various
Chelates to Apotransferrin

Chelates	k_{obs}
Fe-EDTA[a]	0.85×10^{-3} min^{-1}
Fe-citrate	2.5×10^{-3} min^{-1}
Fe-AHA	1.59×10^{-2} sec^{-1}
Fe-GHA	0.17 sec^{-1}
Fe-NTA[b]	0.55 sec^{-1}

[a]Ethylenediaminetetraacetate
[b]Nitrilotriacetate
Source: Ref. 154.

It has been found recently that fructose is especially effective in moving Fe(III) across membranes [155] and also in promoting absorption of orally administered iron [156]. At high concentrations, fructose permits the hydrolytic polymerization of iron(III) in neutral solution, but the resulting polymers are much smaller than those observed in citrate or noncomplexing media and dissociate fairly readily. Fructose also reduces iron(III) to iron(II), but the rate of reduction is much slower than the rate of polymer dissociation.

In a similar manner, various criteria have been suggested for the design of suitable ligands for effective iron chelation in biological systems [148]. These include (1) iron binding capacity; (2) bioavailability and biocompatability. Schubert has stressed that the in vivo binding capacity of a chelate may be substantially reduced by other ions (e.g., Ca^{2+}), and to combat this the extra stability of the chelate effect of multidentate ligands may be employed [157]. With regard to bioavailability and biocompatability, passage through the gastrointestinal epithelium is promoted by increasing the lipophilicity of the drug., e.g., by introducing long-chain alkyl substituents.

Finally, it is essential, in the design of both iron chelates as iron-containing drugs and of chelating ligands as drugs for treatment of iron overload, that the final preparation not have any undesirable side effects in the patient. Regretably, this point can only be proved unequivocally by lengthy testing.

GLOSSARY OF MEDICAL TERMS

Angina: Sore throat from any cause

(Pure red cell) Aplasia: Disorders characterized by isolated loss of erythroid precursors in the bone marrow

Dyspnea: Breathlessness or distressed breathing

Erythropoiesis: Process of formation of red blood cells in the bone marrow

Hemachromatosis: A condition resulting from increased iron absorption for a long period of time.

Hereditary hemorrhagic telangiectasia: A congenital dilatation and thinning out of walls of certain small blood vessels, hence spontaneous rupture and bleeding from the nose, into skin, from gastrointestinal tract

Hyperpnea: A condition in which the respiration is deeper and more rapid than normal.

Hypochromia: An anemic condition in which the percentage of hemoglobin in the red cell is less than the normal range, and in which purpura red corpuscles, with bioconcave shape, may appear more pale in the center than at the periphery

Hypoxia: A general term denoting acute or chronic oxygen deficiency in the tissue from any cause in which oxygen-carrying capacity of the blood is reduced

Idopathic steatomahea: Denoting disease of unknown cause; sebaceous tumor

MCHC: Mean corpuscular hemoglobin concentration

MCV: Mean corpuscular volume

Megaloblastic anemia: Results from cause interfering with the syn-
    thesis of nucleoproteins essential for mitosis of developing
    red cells

Necrosis: The pathologic death of one or more cells or portion of
    tissues from irreversible damage

Neoplastic infiltration: The act of interpenetrating a substance or
    passing into cells or tissues, pertaining to neoplasia

NF: *Normal Formulary*

Normoblast: A nucleated red blood cell; the immediate precursor of
    a normal erythocyte in humans

Parasitism: Infestation with parasites; the mode of existence of
    parasites

Phlebetomy therapy: Bleeding by vein puncture

Polychromatophilic cells: Cells characterized by a greater affinity
    for basic stains--as in a Wright's stained smear, where the
    reticulocyte is slightly larger than a mature erythrocyte and
    is identified by a slight diffuse basophillia that is termed
    polychromatophilia

Reticulocyte: A young red blood cell with a network of precipitated
    basophilic substance and occurring during the process of
    active blood generation

Sickle cells: A condition in which hemoglobin has the property of
    forming crescent-shaped tactoids in the reduced state and in
    which red cells thus assume a sickled form

Sideroblasts: An erythrocyte containing granules of ferritin stained
    by Prussian blue reaction

Sphrecytic anemia: An inherited chronic disease characterized by
    hemolysis of spherical red blood cells

Syncope: Simple form of shock (fainting)

Tachycardia: Distressed circulation systems

Thrombocytopenic bleeding: Purpura reduction in the quality of
    platelets

Telangiectasia: Dilatation of the previously existing small or
    terminal vessels of a part

Ulcerative colitis: Inflammatory disease of the colon, often accom-
      panied by systemic symptoms and characterized by remissions
      and relapses

USP: *United States Pharmacopoeia*

REFERENCES

1. T. G. Spiro and P. Saltman, *Struct. Bonding, 6,* 116 (1969).

2. D. L. Leussing and J. M. Kolthoff, *J. Amer. Chem. Soc., 75,*
    2476 (1953).

3. L. G. Sillén and A. E. Martell, *Stability Constants,* Special
    Publication 17, Chemical Society, London (1964).

4. G. Bierdermann, *Prog. Inorg. Chem., 9,* 1 (1968).

5. B. O. A. Hedström, *Ark. Kemi, 6,* 1 (1953).

6. R. M. Milburn and W. C. Vosburgh, *J. Amer. Chem. Soc., 77,*
    1352 (1955).

7. G. Bierdermann, quoted in K. Schyter, *Trans. Roy. Inst.*
    Technol. Stockholm, No. 196 (1962).

8. H. B. Gray, in *Proteins of Iron Storage and Transport in Bio-*
    chemistry and Medicine (R. R. Crichton, ed.), North-Holland/
    Elsevier, New York, 1975.

9. S. J. Lippard, H. Schugar, and C. Walling, *Inorg. Chem., 6,*
    1825 (1967).

10. H. Schugar, G. R. Rossman, and H. B. Gray, *J. Amer. Chem. Soc.,*
    91, 4564 (1969).

11. G. Anderegg, *Helv. Chim. Acta, 43,* 1530 (1960).

12. B. A. Sommer and D. W. Margerum, *Inorg. Chem., 9,* 2517 (1970).

13. H. N. Po and N. Sutin, *Inorg. Chem., 10,* 428 (1971).

14. T. G. Spiro, S. E. Allerton, J. Denner, A. Terzis, R. Bils, and
    P. Saltman, *J. Amer. Chem. Soc., 88,* 2721 (1961).

15. G. W. Brady, C. R. Kurkjian, E. F. X. Lyden, M. B. Robin, P.
    Saltman, T. G. Spiro, and A. Terzis, *Biochemistry, 1,* 2105
    (1968).

16. H. B. Gray, in *Bioinorganic Chemistry* (R. F. Gould, ed.), Ad-
    vances in Chemistry Series No. 100, American Chemical Society,
    Washington, D.C., 1971, p. 365.

17. T. G. Spiro, L. Pape, and P. Saltman, *J. Amer. Chem. Soc., 89,*
    5555 (1967).

18. T. G. Spiro, G. Bates, and P. Saltman, *J. Amer. Chem. Soc., 89,* 5559 (1967).

19. A. Simmons, *Basic Hematology,* Charles C. Thomas, Springfield, Ill., 1973, p. 6.

20. S. Granick, Iron metabolism in animals and plants, in *Trace Elements,* Academic Press, New York, 1958, p. 365.

21. E. Bentler and S. K. Srivastava, in *Hematology* (W. Williams, ed.), McGraw-Hill, New York, 1972, pp. 94-100.

22. J. F. Seitz, *The Biochemistry of the Cells of Blood and Bone Marrow,* Charles C. Thomas, Springfield, Ill., 1969, p. 10.

23. H. N. Antoniades, *Hormones in Human Blood,* Harvard University Press, Cambridge, 1976, p. 19.

24. N. Dimitrov and R. L. S. Jernholm, *Blood, 28,* 998 (1966).

25. O. Siggard-Anderson, *The Acid-Base Status of Blood,* Williams & Wilkins, Baltimore, 1974, p. 13.

26. E. Hackett, *Blood,* Saturday Review Press, New York, 1973, p. 30.

27. C. G. Holmberg and C. B. Laurell, *Acta Chem. Scand., 1,* 944 (1947).

28. J. H. Jandl and J. H. Katz, *J. Clin. Invest., 42,* 314 (1963).

29. E. H. Morgan and C. B. Laurell, *Brit. J. Haemat., 9,* 471 (1963).

30. A. L. Schade, *Nutr. Rev., 13,* 225 (1955).

31. J. H. Jandl, J. K. Inman, R. L. Simmons, and D. W. Allen, *J. Clin. Invest., 38,* 161 (1959).

32. A. Mazur, S. Green, and A. Charleton, *J. Biol. Chem., 235,* 395 (1960).

33. L. E. Strong, W. L. Hughes, Jr., D. J. Mulford, J. H. Asworth, M. Melin, and H. L. Taylor, *J. Amer. Chem. Soc., 68,* 459 (1946).

34. A. L. Schade and B. L. Carlominen, *Science, 104,* 340 (1946).

35. E. J. Cohn, *Experientia, 3,* 125 (1947).

36. D. M. Surgenor, B. A. Koechlin, and L. E. Strong, *J. Clin. Invest., 28,* 73 (1949).

37. D. M. Surgenor, H. L. Taylor, R. S. Gordon, Jr., and D. M. Gibson, *J. Amer. Chem. Soc., 71,* 1223 (1949).

38. B. A. Koechlin, *J. Amer. Chem. Soc., 74,* 2649 (1952).

39. C. B. Laurell and B. Ingelman, *Acta Chem. Scand., 1,* 770 (1947).

40. G. A. Jamieson, *J. Biol. Chem., 240,* 2914 (1965).

41. J. J. Windle, A. K. Wiersema, J. R. Clark, and R. E. Feeney, *Biochemistry, 2,* 1341 (1963).

42. P. R. Azari and R. E. Feeney, *J. Biol. Chem., 232,* 293 (1958).

43. P. R. Azari and R. E. Feeney, *Arch. Biochem. Biophys.*, *92*, 44 (1961).

44. A. N. Glazer and H. A. McKenzie, *Biochem. Biophys. Acta*, *71*, 109 (1963).

45. S. Fiala and D. Burk, *Arch. Biochem.*, *20*, 172 (1949).

46. R. C. Warner and I. Weber, *J. Biol. Chem.*, *191*, 173 (1951).

47. B. Davis, P. Saltman, and S. Benson, *Biochem. Biophys. Res. Commun.*, *8*, 56 (1962).

48. R. Aasa, B. G. Malmström, P. Saltman, and T. Vänngård, *Biochim. Biophys. Acta*, *75*, 203 (1963).

49. J. O. Jeppsson and J. Sjoberg, 6th Int. Cong. Biochem., II, 157 (1964).

50. S. T. Arend, M. L. Gallango, W. C. Parker, and A. G. Bearn, *Nature*, *196*, 447 (1962).

51. Y. C. Lail, *Nature*, *198*, 589 (1963).

52. R. F. Murray, Jr., J. C. Robinson, and B. S. Blumberg, *Nature*, *204*, 382 (1964).

53. W. C. Parker and A. G. Bearn, *Science*, *137*, 854 (1962).

54. N. Schmiedeberg, *Arch. Exp. Pathol. Pharmakol.*, *33*, 101 (1894).

55. V. Laufberger, *Biol. Listy.*, *19*, 73 (1934).

56. S. Granick, *J. Biol. Chem.*, *149*, 157 (1943).

57. V. Laufberger, *Bull. Soc. Chem. Biol.*, *19*, 1575 (1937).

58. A. Mazur and E. Shorr, *J. Biol. Chem.*, *176*, 771 (1948).

59. R. Kuhn, N. A. Sörenson, and L. Birkhofer, *Ber. Chem. Ges.*, *73B*, 823 (1940).

60. S. Granick, *J. Biol. Chem.*, *146*, 451 (1942).

61. A. Mazur, I. Litt, and E. Shorr, *J. Biol. Chem.*, *187*, 473 (1950).

62. M. C. Linder and H. N. Munro, *Anal. Biochem.*, *48*, 266 (1972).

63. J. L. Farrant, *Biochim. Biophys. Acta*, *13*, 569 (1954).

64. P. M. Harrison, *J. Mol. Biol.*, *6*, 404 (1963).

65. H. I. Bielig, O. Kratky, G. Rohns, and H. Wawra, *Biochim. Biophys. Acta*, *112*, 110 (1966).

66. V. Kleinwachter, *Arch. Biochem. Biophys.*, *105*, 352 (1964).

67. F. A. Fischbach and J. W. Andergg, *J. Mol. Biol.*, *14*, 458 (1965).

68. F. Friedberg, *Can. J. Biochem. Physiol.*, *40*, 983 (1962).

69. P. M. Harrison, T. G. Hoy, I. G. Macara, and R. J. Hoare, *Biochem. J.*, *143*, 445 (1974).

70. L. Michaelis, C. D. Coryell, and S. Granick, *J. Biol. Chem.*, *148*, 463 (1943).

71. S. Granick and P. F. Hahn, *J. Biol. Chem.*, *155*, 661 (1944).

72. P. M. Harrison and T. Hofmann, *J. Mol. Biol.*, *4*, 239 (1962).

73. T. Hofmann and P. M. Harrison, *J. Mol. Biol.*, *6*, 256 (1963).

74. P. M. Harrison, T. Hofmann, and W. I. P. Mainwaring, *J. Mol. Biol.*, *4*, 251 (1962).

75. W. I. P. Mainwaring and T. Hofmann, *Arch. Biochem. Biophys.*, *125*, 975 (1968).

76. R. R. Crichton and F. C. A. Bryce, *FEBS Lett.*, *6*, 121 (1970).

77. A. Mazur, S. Baez, and E. Shorr, *J. Biol. Chem.*, *213*, 147 (1955).

78. H. J. Bielig and E. Bayer, *Naturwissenschaften*, *42*, 466 (1955).

79. S. Green and A. Mazur, *J. Biol. Chem.*, *277*, 653 (1967).

80. A. Mazur, S. Green, and A. Saha, *J. Clin. Invest.*, *37*, 1809 (1958).

81. J. A. Smith, J. W. Drysdale, A. Goldberg, and H. N. Munro, *Brit. J. Haemat.*, *14*, 79 (1968).

82. G. M. Brittin and D. Raval, *J. Lab. Clin. Med.*, *75*, 811 (1970).

83. M. Linder-Horowitz, R. T. Reutinger, and H. N. Munro, *Biochim. Biophys. Acta*, *200*, 442 (1970).

84. M. Linder, H. N. Munro, and H. P. Morris, *Cancer Res.*, *30*, 2231 (1970).

85. J. B. Neilands, *Struc. Bonding*, *1*, 100 (1966).

86. H. J. Yale, The hydroxamic acids, *Chem. Rev.*, *33*, 209 (1943).

87. F. Mathis, The hydroxamic acids, *Bull. Soc. Chim. Fr.*, D9-D22 (1953).

88. D. A. Brown, D. McKeith, and W. K. Glass, *Inorg. Chim. Acta*, *35*, 5, 57 (1979).

89. A. H. Blatt, *Organic Syntheses, Coll.*, Vol. 2, Wiley, New York, 1943, p. 69.

90. R. L. Dutta, *J. Ind. Chem. Soc.*, *36*, 385 (1959).

91. S. Seifter, P. M. Gallup, S. Michaels, and E. Meilman, *J. Biol. Chem.*, *235*, 2613 (1960).

92. T. Emery, *J. Amer. Chem. Soc.*, *83*, 1626 (1961).

93. E. A. Kaczka, C. O. Gitterman, E. L. Dulaney, and K. Folkers, *Biochemistry*, *1*, 340 (1962).

94. T. Emery and J. B. Neilands, *Nature*, *184*, 1632 (1959).

95. W. Keller-Schierlein and H. Zahner, *Progr. Chem. Org. Nat. Prod.*, *22*, 279 (1964).

96.   J. B. Neilands, *Experientia, Suppl., IX,* 22 (1964).

97.   A. Ricicova, *Collec. Czech. Chem. Commun., 28,* 1761 (1963).

98.   H. Zahner, E. Bachmann, R. Hutter, and J. Neusch, *Pathol. Microbiol., 25,* 708 (1962).

99.   H. Keberle, *Ann N.Y. Acad. Sci., 119,* 758 (1964).

100. *Time,* Oct. 9, 1964, p. 66.

101. A. Zalkin, J. D. Forrester, and D. H. Templeton, *Science, 146,* 261 (1964).

102. A. Ehrenberg, *Nature, 178,* 379 (1956).

103. V. Prelog, *Pure Appl. Chem., 6,* 327 (1963).

104. Committee on Iron Deficiency, AMA, Council on Food and Nutrition, Iron deficiency in the United States, *J. Amer. Med. Ass., 203,* 407 (1968).

105. H. S. White, Iron nutrition of girls and women: A review, *Amer. Dietet. Ass., 53,* 563-574 (1968).

106. World Health Organization Scientific Group on Nutritional Anemias, Nutritional Anemias World Health Organ. Tech. Report, Ser. No. 405, 1-37 (1968).

107. Food and Nutrition Board, *Recommended Dietary Allowances,* 7th ed., National Academcy of Sciences, Washington, D.C., 1968, p. 1694.

108. E. J. Underwood, *Trace Elements in Human and Animal Nutrition,* Academic Press, New York and London, 1962, Chap. 2.

109. E. R. Monsen, I. N. Kuhn, and C. A. Finch, Iron status of menstruating women, *Amer. J. Clin. Nutr., 20,* 842, 849 (1967).

110. Committee on Iron Deficiency, AMA, Council on Foods and Nutrition, Iron deficiency in the United States, *J. Amer. Med. Ass., 203,* 119-214 (1968).

111. Food and Nutrition Board, *National Research Council Recommended Dietary Allowances,* 8th rev. ed., Publication No. 2216, National Academy of Sciences, Washington, D.C., 1974.

112. FAO/WHO Expert Group, Requirements of ascorbic acid, vitamin D, vitamin B_{12}, folate, and iron, WHO Technical Reports, World Health Organization, Geneva, 1970.

113. E. Underitz, in *Iron Metabolism* (F. Gross, ed.), Springer, Berlin, 1964, p. 407.

114. H. Brise and L. Hallberg, *Acta Med. Scand., 171, Suppl.,* 376, 23-38 (1962).

115. P. V. Sacks and W. H. Grosby, *Clin. Res., 22,* 562A (1974).

116. V. F. Fairbanks, J. L. Fahey, and E. Bentler, *Clinical Disorders of Iron Metabolism,* Grune & Stratton, New York, 1971.

117. R. D. Wallerstein and S. R. Mettier, eds., *Iron in Clinical Medicine*, University of California Press, Berkeley, 1958.

118. T. H. Bothwell and C. A. Finch, *Iron Metabolism*, Little, Brown, Boston, 1962.

119. S. Moeschlin and U. Schneider, in *Iron Metabolism* (F. Gross, ed.), Springer, Berlin, 1964.

120. R. H. Girdwood, *Brit. Med. J.*, *1*, 599 (1952).

121. L. Hallberg, H. G. Harworth, and A. Vannotti, eds., *Iron Deficiency*, Academic Press, New York, 1970.

122. J. A. Nissam, *Brit. Med. J.*, *1*, 352 (1954).

123. K. J. Agner, N. S. E. Anderson, and N. G. Nordenson, *Acta Haemat.* (Switz.), *1*, 193 (1948).

124. J. Badenoch and S. T. Callender, *Blood*, *9*, 123 (1954).

125. C. F. Hawkins, A. L. P. Peeney, and W. T. Cooke, *Lancet, II*, 387 (1950).

126. N. S. E. Andersson, *Acta Med. Scand.*, *Suppl.*, *241*, 5 (1950).

127. J. E. Lucas and A. B. Hagedorn, *Blood*, *7*, 358 (1952).

128. J. Fielding, *Brit. Med. J.*, *II*, 279 (1961).

129. W. G. Figueroa, in *Iron Metabolism* (F. Gross, ed.), Springer, Berlin, 1964, p. 429.

130. L. Goldberg, in *Iron in Clinical Medicine*, University of California Press, Berkeley and Los Angeles, 1958, p. 74.

131. L. Garby and S. Sjolin, *Acta Chem. Scand.*, *157*, 319 (1957).

132. A. J. Grimes and M. S. R. Hutt, *Brit. Med. J.*, *II*, 1074 (1957).

133. N. S. E. Andersson, *Brit. Med. J.*, *II*, 275 (1961).

134. S. Lindvall and N. S. E. Andersson, *Brit. J. Pharmacol.*, *17*, 358 (1961).

135. N. S. E. Andersson, *Brit. Med. J.*, *II*, 1260 (1962).

136. A. A. Pringle, A. Godberg, E. McDonald, and S. Johnston, *Lancet, II*, 749 (1962).

137. T. Bersin and H. Schwarz, *Justus Liebigs Ann. Chem.*, *623*, 138 (1959).

138. T. Mereu and O. Tonz, *Dtsch. Med. Wochenschr.*, *86*, 1259 (1961).

139. E. B. Brown and C. V. Moore, *Progr. Haemat.*, *1*, 22 (1936).

140. D. H. Coleman, A. R. Stevens, and C. A. Finch, *Blood*, *10*, 567 (1955).

141. R. K. Bass, E. R. Halden, and E. E. Muirhead, *Texas State J. Med.*, *55*, 22 (1959).

142. J. M. Scott and A. D. T. Govan, *Brit. Med. J., II,* 1257 (1954).

143. H. G. Richmond, *Scott. Med. J., 2,* 169 (1957).

144. H. G. Richmond, *Brit. Med. J., I,* 947 (1959).

145. A. Haddow and E. S. Horning, *J. Nat. Cancer Inst., 24,* 109 (1960).

146. A. S. Blazar and M. G. Delkiego, *Obster. Gyner., 20,* 145 (1962).

147. H. S. Waxman and E. B. Brown, *Progr. Haemat., 6,* 338 (1969).

148. W. F. Anderson and M. C. Hiller, eds., *Development of Iron Chelators for Clinical Use,* U.S. DHEW Pub. No. (NIH) 76-994, 1975.

149. R. W. Grady, J. H. Graziano, G. P. White, A. Jacobs, and A. Cerami, *J. Pharmacol. Exp. Ther., 205,* 757 (1978).

150. D. A. Brown, M. V. Chidambaram, J. J. Clarke, and D. M. McAleese, *Bioinorg. Chem., 9,* 255 (1978).

151. D. A. Brown, M. V. Chidambaram, and J. D. Glennon, *Inorg. Chem., 19,* 3260 (1980).

152. F. Basolo and R. G. Pearson, *Mechanisms of Inorganic Reactions,* Interscience, New York, 1958.

153. G. W. Bates and J. Wernicke, *J. Biol. Chem., 246,* 3679 (1971).

154. J. D. Glennon, Ph.D. thesis, National University of Ireland, p. 145, 1979.

155. C. Stitt, P. Charley, E. Butt, and P. Saltman, *Proc. Soc. Exp. Biol. Med., 10,* 70 (1962).

156. G. L. Bates, J. Boyer, J. C. Hegenauer, and P. Saltman, *Amer. J. Clin. Nutr., 25,* 983 (1972).

157. J. Schubert, The chemical basis of chelation, in *Iron Metabolism* (F. Gross, ed.), Springer, Berlin, 1964.

158. B. A. Brown, *Hematology: Principles and Procedures,* 2nd ed., Lee and Febiger, Philadelphia, 1976.

159. M. Wintrobe, *Clinical Hematology,* 6th ed., Lee and Febiger, Philadelphia, 1967.

Chapter 6

GOLD COMPLEXES AS METALLO-DRUGS

Kailash C. Dash[*] and Hubert Schmidbaur
Institute of Inorganic Chemistry
Technical University of Munich
Garching, Federal Republic of Germany

---

[*]Present affiliation: Department of Chemistry, Utkal University, Vani Vihar, Bhubaneswar, India.

1. INTRODUCTION

Even though the empirical use of gold in medicine is recorded far
back in history, modern interest in its medicinal use may be traced
to the stimulating work by Koch [1], which put gold therapy (*chryso-*
therapy) on a sound basis and paved the way for the use of gold
salts for the treatment of tuberculosis, syphilis, and a number of
other pathogenic organisms. Gold thioglucose--one of the gold drugs
still in use today--was first tested for the treatment of tuberculosis by Feldt in 1917 [2]. It was not until 1927, however, that
Landé [3], while employing gold thioglucose (GTG) in the treatment
of bacterial endocarditis, observed that joint pain was relieved in
his patients and suggested the beneficial effects of GTG in arthritis. Almost at the same time, Forestier [4] also noted similar observations, and thus active interest in chrysotherapy was created.
Gold sodium thiosulfate, another gold complex still in use, was mentioned by Möllgaard also as a drug for tuberculosis as early as 1924
[5]. Though at the present time gold compounds are used almost exclusively against arthritis and related conditions, we will also
briefly consider here their effects in altering the progression of
other diseases.

Topical reviews on this subject, each with a different emphasis, have been published by Sadler [6], Shaw [7], and Brown and
Smith [8].

1.1. Asthma

Gold salts are of value in the treatment of asthma of tuberculosis
origin [9], and the gold cation has an affinity for the lung tissue
[10]. Gold-treated guinea pigs develop allergic asthma, however
[11]. Gold-sulfur complexes have found use against bronchial asthma
[12].

1.2. Tuberculosis

Sodium aurosulfite, $Na_5Au(SO_3)_4 \cdot 5H_2O$, was reported to be an important drug against tuberculosis [13]. GTG (50% Au) and gold calcium keratinate [14% Au) were also believed to be active against the human tubercle bacillus [14].

1.3. *Pemphigus Vulgaris*

This skin disease is due to the autoimmune disorder of the skin and can be fatal if neglected. Both corticosteroids and gold compounds are used for its treatment, but gold drugs appear to be preferred [15,16].

1.4. Antimicrobial

In vitro tests using gold complexes with certain sulfur drugs (e.g., sulfadiazine, sulfamethizole, and sulfasomidine) against a wide variety of bacteria show a higher antimicrobial activity as compared to the parent sulfur compounds [16]. The gold sulfamides (Structure I) are bactericides [16].

$$H_2N-\langle\bigcirc\rangle-SO_2\text{-}\underset{\underset{Au}{|}}{N}-R$$

(I)   R = heterocyclic radical

Gold sulfadiazine prevents the growth of *Bacillus subtilis*, *E. coli*, *Salmonella typhii*, *Shigella dysenteria*, *P. vulgaris*, *Proteus morganii*, *Klebisiella pneumonia*, etc. Several gold compounds of some common sulfur-containing drugs are useful for the treatment of ringworms and fungal infections [17,18].

1.5. Antitumor

Gold is apparently not carcinogenic, and a few gold compounds are active against cancer. The T/C (survival time of treated animals/ survival time of controls) of the gold complexes--e.g., $[AuCl_4][HL]$ (L = antipyrine derivatives)--indicated antitumor activities [19]. In therapeutic concentrations, gold thiopolypeptides decrease cell proliferation and glycosamin-glycan formation and are effective against tissue-forming cells [20]. 5-Substituted tetrazole tri- phenylphosphine gold(I) complexes (II) prepared according to Eq. (1) are found to have antiarthritic and antitumor activities [21].

$$Ph_3PAu(OAc) + HN_4CR \longrightarrow Ph_3PAu(N_4CR) + HOAc \qquad (1)$$

$$HN_4CR = \underset{\substack{N-N}}{\overset{R}{\underset{N}{\overset{C}{\diagup}}}} - H \qquad \begin{array}{l} R = NMe_2, H, Ph, \\ NH_2, CF_3 \end{array}$$

(II)

Therefore, a significant pharmacological role for gold seems in hand, particularly in view of the fact that the complex $[Au(5\text{-diazouracil})_2 Cl_2]Cl \cdot HCl$ also has antitumor activities (in mice) [22]. 2,3,4,6- Tetra-O-acetyl-1-thio-β-D-glucopyranosido-S(triethylphosphine) gold, auranofin (vide infra), apparently exerts a significant inhibitory effect on biological processes and functions essential to cell pro- liferation, suggesting its antitumor capabilities [23]. Auranofin inhibits DNA synthesis by HeLa cells and suppresses [3H]thymidine uptake more readily than [3H]leucine or [3H]uridine uptake. Admin- istration of colloidal [198]Au protected rats against hepatic tumor growth [24].

1.6. Rheumatoid Arthritis

Arthritis is an autoimmune disease which is imperfectly understood. It is postulated that impaired sulfhydryl group reactivity due to accelerated formation of disulfide bonds results in macroglobulin formation and protein denaturation, thus triggering the immune system. Sulfhydryl binding agents are capable of suppressing the inflammatory process, and gold acts in a similar way [25,26]. Gold finds its major therapeutic use today in the treatment of rheumatoid arthritis, predominantly with sodium aurothiomalate and aurothioglucose, and to a lesser extent aurothiosulfate and aurothiopolypeptide. The efficacy of gold therapy in the treatment of rheumatoid arthritis has been established in a series of controlled trials [27-29], and gold is one of the few drugs that can alter the course of this disease. However, although some patients can safely remain on gold therapy for many years, sometimes toxic side effects do occur [30, 31].

In what follows, a greater emphasis will be laid on the use of gold complexes against arthritic diseases.

2. GOLD DRUGS AGAINST ARTHRITIS

Rheumatoid arthritis is a chronic inflammatory disease characterized by lymphoid infiltration, granuloma formation, and an imbalance of humoral and cellular immune mechanisms. A number of clinical and biological assessments are used to measure the disease activity, and the clinical improvement is accompanied by biological modifications, e.g., reductions of erythrocyte sedimentation rate, of immune complex concentrations, of urinary excretions of the lysosomal enzyme β-N-acetyl glucosaminidase, etc., and increase in the hemoglobin concentrations. The biological modifications during the drug treatment should tend toward normalization and indicate reduced inflammatory activity.

2.1. Parenterally Administered Drugs

The most important *gold-free* drugs known to be effective against arthritis are fenoprofen (FPF) (III), tolmetin (IV), levamisole (V), penicillamine (VI), aspirin, and the corticosteroid halcinonide.

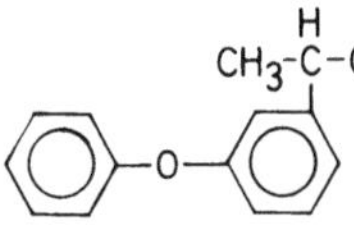

(III)

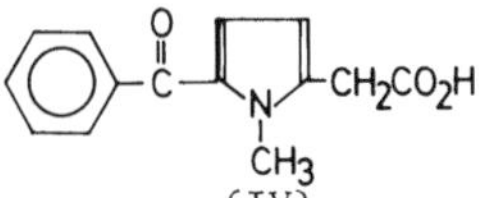

(IV)

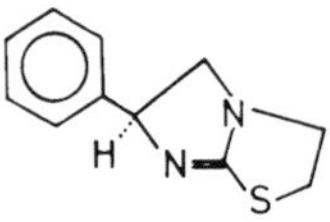

(V)

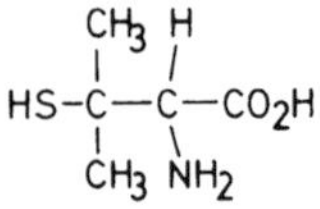

(VI)

In addition, many *gold* complexes--mainly gold(I) thiolates--
are also effective. Until very recently these have been usually administered parenterally, i.e., through intramuscular injection.

2.1.1. *Some Early Gold Drugs*

On the basis of his studies on the pharmacological action of $AuCl_3$,
$Na_3[Au(S_2O_3)_2]$, colloidal gold sulfide, and colloidal gold, Orestano
[32,33] suggested in the early thirties that in each of these cases
the substances are transformed in the organisms into granular gold,
which is the least active state from a physical or chemical stand-
point. Though all these compounds failed to influence the course of
experimental tuberculosis of rabbits, they were pharmacologically
quite active. Orestano also pointed out that $Na_3[Au(S_2O_3)_2]$ was absorbed without any toxic effects characteristic of $AuCl_3$, but produced
violent diarrhea and nephrosis.

The insoluble calcium aurothioglycolate (Calaurol, Myoral) was
later found to be much safer than some other gold drugs, but at the
same time was also less effective [34]. Gold thiomalic acid, $HOOCCH_2$-
$CH(CH_2SAu)COOH$, and its Na and Ca salts were also described as useful
against rheumatoid arthritis and lupus erythematosus [35]. In a subsequent series of experiments, Block et al. [36-38] studied the

effect of gold calcium thiomalate, sodium succinimidoaurate, gold
sodium thiomalate, gold sodium thiosulfate, and colloidal "Au$_2$S$_3$" in
rheumatoid arthritic patients with active joint synovitis. They
found much gold in the liver and kidneys, considerably less in the
spleen, and very little in the other organ tissues. The metal was
excreted in both urine and feces.

It is important to note that none of the above gold compounds
is structurally fully characterized. Their formulas were proposed
solely from analytical data and the bonding of gold to sulfur as-
signed only intuitively.

2.1.2. *Gold Trisodium Thiosulfate,*
        Na$_3$[Au(S$_2$O$_3$)$_2$]

Gold sodium thiosulfate, Na$_3$[Au(S$_2$O$_3$)$_2$], known clinically as Sanocry-
sin, has been used for the treatment of arthritis, tuberculosis, and
leprosy [32,36]. It can be prepared by reducing HAuCl$_4$ with Na$_2$S$_2$O$_3$
[39]. The x-ray crystal structure [40,41] shows the Au atom bonded
to two S atoms from two thiosulfate groups in a nearly linear arrangement [S-Au-S angle is 176.5(2)°] (VII). The drug is administered

(VII)

Au-S$_3$	2.28 Å	
Au-S$_4$	2.27	
S$_1$-S$_4$	2.05	
S$_2$-S$_3$	2.07	
S$_4$AuS$_3$	176.5°	
AuS$_3$S$_2$	103.6	
AuS$_4$S$_1$	104.1	

in aqueous solution, but in contrast to most other classical gold
drugs, intravenous injection is also possible. (All others are administered intramuscularly [42].)

2.1.3. Gold Disodium Thiomalate (tm)

Gold disodium thiomalate, $Na_2Au(tm)$ (VIII), known clinically as

$$Au-S-\underset{\underset{CH_2-CO_2Na}{|}}{CH}-CO_2Na$$

(VIII)

Myocrisin, has survived the test of time in its use against arthritis. It can be prepared from thiomalatogold(I), $Au(tmH_2)$, and NaOH
in water [43]. From recent infrared spectroscopic data it appears
that the compound contains an Au-S bond and that the carboxylate
groups do not complex strongly, if at all, with gold [43]. In
another study [44] the existence of a cluster consisting of four gold
atoms and seven thiomalate anions in aqueous solution was derived
from molecular weight and ^{13}C nuclear magnetic resonance measurements.

An oil suspension of this compound gives favorable results
against chronic rheumatism. Administered parenterally, it suppresses both the primary and secondary lesions while increasing the
Au levels and is effective against adjuvant-induced arthritis [45-
48]. $Na_2Au(tm)$ inhibits both the humoral and cellular immune mechanisms and appears to act primarily at the accessory (macrophage)
cell level with a possible secondary effect on T-lymphocytes [49].
It inhibits the DNA synthesis in sheep lymphocytes cultured in the
presence of phytohemaglutinin (PHA) [50]. Like other anti-inflammatory drugs, it exerts its inhibitory effect by a stabilizing action
on lysosomal membranes [51]. Apparently, $Na_2Au(tm)$ breaks down in
the cell to release the Au^+ cation, which is a strong noncompetitive
inhibitor for sulfhydryl-containing enzymes [52]. This inhibition,
however, is reversed by the addition of a sulfhydryl compound, e.g.,
cysteine [52].

2.1.4. Gold Thioglucose

Gold thioglucose (GTG) (IX), known clinically as Solganol, is widely

(IX)

used against arthritis. It may be prepared from an alcoholic solution of $HAuBr_4$, reduced by passing SO_2 gas, and a solution of thioglucose [53]. It is administered parenterally and because of its
higher molecular weight of approximately 1000 (due to polymerization)
seems to be less toxic than $Na_2Au(tm)$. GTG induces hyperphagia (increased macrophage level), insatiable appetite, and obesity in certain strains of mice [54], primarily due to a metabolic disorder [55].
A number of gold thio compounds do not induce obesity and the thioglucose moiety seems to act as a specific carrier of the gold [56].
However, the obesity need not be of great concern as it is not observed in other species (rats and humans), and besides, a heavy dose
(350-500 mg Au/kg body weight) is necessary to induce obesity as compared to a typical clinical dose in chrysotherapy (0.35 mg Au/kg).
The use of GTG as an antiarthritic compound is thus still significant.
GTG is administered as an oil suspension intramuscularly, and a depot
effect is noticeable.

2.1.5. Gold Thiopolypeptide

Certain fractions of hydrolyzed protein are known to form well-defined
gold(I) complexes containing approximately 13% Au. Human hair protein
and other starting materials are used in the preparation. Radioactive
[198]Au has also been incorporated into the complexes. Aqueous solutions are commercial drugs (Aurodetoxin) [42].

2.2. Orally Administered Drugs

The most common form of chrysotherapy (v.i.) requires intramuscular
injections of the gold preparation, usually of the order of 50 mg,
at weekly intervals, but since gold retention and toxicity may persist for long periods of time after treatment has ceased, it is important to administer the smallest amount of gold drug for efficient
therapeutic effects. It is possible that a gold drug that could be
absorbed orally would be of great value in permitting the administration of minimum daily doses and of facilitating the control of
both dosage and related toxicity. The known soluble gold drugs
given orally provide no protection against adjuvant arthritis [57,
58], and the need for a new oral chrysotherapeutic agent was obvious.

2.2.1. Gold(I) Phosphines and Related
Complexes as Oral Drugs

In searching for new oral chrysotherapeutic agents, Sutton et al. [59]
noted that alkylphosphine gold complexes, R_3PAuX, administered orally
were as effective as other anti-inflammatory agents for adjuvant arthritic rats and were absorbed efficiently as shown by serum Au
levels. The R_3PAuCl complexes, easily prepared from ethanolic solution of $HAuCl_4$ (1 mol) and R_3P (2 mol), or from AuCl and PR_3, are
highly soluble in lipid, and their therapeutic responses as well as
oral absorption properties, as evidenced by serum Au levels, seem to
be structure-dependent. Although the nongold phosphine compounds
gave no protection, comparative studies indicated that the nature of
the phosphine ligand in the gold complexes played a greater role in
bringing about changes in biological activity than did the other
group (X) bonded directly to Au. The serum gold levels, as well as
the therapeutic effect, were highest with the Et_3PAuCl complex, and
with increase in the bulkiness of the alkyl substituent (e.g., on
going to i-Pr or n-Bu) both serum Au levels and therapeutic protection decreased. Oral Et_3PAuCl proved as effective as intramuscular
$Na_2Au(tm)$ [58]. A comparative study of tissue Au levels produced in

guinea pigs after the oral administration of Et_3PAuCl or $Na_2Au(tm)$, or after the injection of $Na_2Au(tm)$, showed that in gastrointestinal tissues, the tissue Au concentration was highest with Et_3PAuCl [60]. Further, Et_3PAuCl (5 mg/kg per day, oral), but not $Na_2Au(tm)$ (5 mg/kg per day, either injection or oral), effectively suppressed kaolin edema in the rat paw and the effect was independent of serum Au level.

In addition, a number of trialkylphosphine or phosphite complexes, R_3PAuCl or $(RO)_3PAuCl$, and a number of thiocyanato complexes, $R_3PAuSCN$ ($R = C_{1-3}$ alkyl, alkoxy, phenyl, or phenoxy), in daily oral doses of 0.5-10 mg (based on Au content) in the form of tablets or capsules are effective against arthritis in rats [61,62].

2.2.2. Gold(I) Phosphine (or Phosphite) Thiolates

A number of gold(I) phosphine thiolates (see also Sec. 2.2.3) have been prepared [see Eqs. (2) and (3)] and tested against adjuvant arthritis in rats, in the search of an effective oral gold drug.

$$R_3P + AuCl \longrightarrow R_3P-Au-Cl \tag{2}$$
$$R_3P-Au-Cl + R'S^- \longrightarrow R_3PAuSR' + Cl^- \tag{3}$$

where

$$R = Me, Et, i\text{-}Pr, n\text{-}Bu$$
$$X = H, Ac.$$
$$Y = O, S.$$

Phosphine (or phosphite) Au(I) complexes of thioethanols, $R_3PAuSCH_2$ $CHR'OR''$, thiomalic acid, $R_3PAuSCH(COOH)CH_2COOH$ [63], thiobenzoic acid, $Et_3PAuSCPh$, and substituted thiophenols, $o\text{-}HOOCC_6H_4SAuPEt_3$ or $o\text{-}H_2NC_6H_4SAuPEt_3$ [64] are all active against arthritis. A number of other thio compounds, e.g., R_3PAuX (X = 2-thiazolinyl, thio-2-benzimidazolyl, and 2-benzoxazolylthio-) have been also formulated as oral tablets for arthritis. Direct reaction of Et_3PAuCl and $HOOCCH_2S(CH_2)_4PEt_2$ (or $HSCH_2CH_2PEt_2$) yielded the large-ring chelate compound of the type shown in Structure X which is formulated into oral tablets [65].

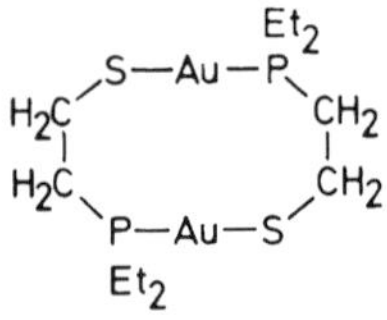

$$
\begin{array}{c}
\text{Et}_2 \\
\text{S—Au—P} \\
\text{H}_2\text{C} \quad \text{CH}_2 \\
\text{H}_2\text{C} \quad \text{CH}_2 \\
\text{P—Au—S} \\
\text{Et}_2
\end{array}
$$

(X)

The molecular structure of this chelate (XI) has been determined recently, and the molecule has a C_2 symmetry with a 10-membered ring structure [66], where the Au, S, and P atoms are not coplanar. In

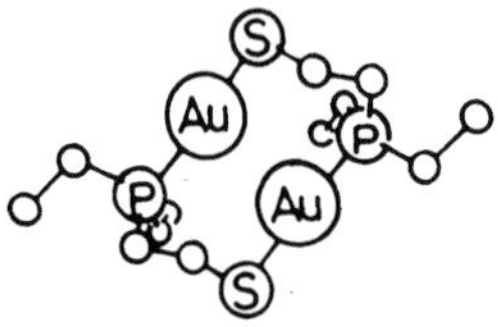

(XI)

Au-Au	3.104 Å	S-Au-P	173.5°
Au-P	2.27	Au-P-C	112.4
Au-S	2.31	Au-S-C	103.8
P-C	1.79	P-C-C	115.8
S-C	1.83	S-C-C	113.7
C-C	1.54		

attempts to prepare a similar dimer by treating the μ-[1,2-bis(diphenylphosphine)ethane]-bis-chlorogold(I) with the dianion of 1,2-ethanedithiol, a polymeric compound (XII) was formed [65].

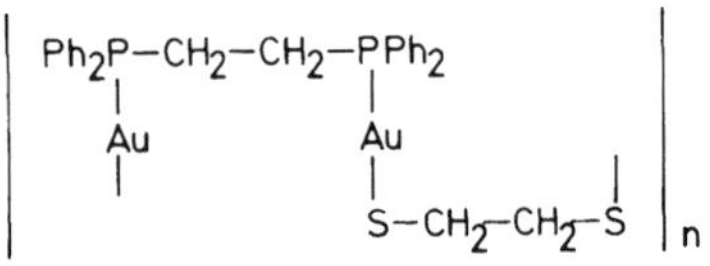

$$
\begin{array}{c}
\text{Ph}_2\text{P—CH}_2\text{—CH}_2\text{—PPh}_2 \\
| \qquad\qquad | \\
\text{Au} \qquad\quad \text{Au} \\
| \qquad\qquad | \\
\text{S—CH}_2\text{—CH}_2\text{—S}
\end{array} \Bigg]_n
$$

(XII)

2.2.3. *Auranofin--A Novel Oral Gold Drug*

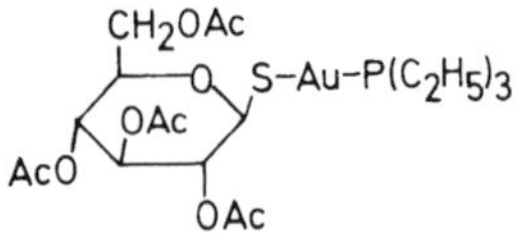

$$
\begin{array}{c}
\text{CH}_2\text{OAc} \\
\text{O} \quad \text{S—Au—P(C}_2\text{H}_5\text{)}_3 \\
\text{OAc} \\
\text{AcO} \\
\text{OAc}
\end{array}
$$

(XIII)

Chemically, auranofin (XIII) is (2,3,4,6-tetra-O-acetyl-1-thio-β-D-glucopyranosido-S)triethylphosphine gold(I) and belongs to a class

of compounds which are trialkylphosphine gold(I) derivatives of glu-
copyranosides. Such compounds are prepared either by treating bis(2,
3,4,6-tetra-O-acetyl-β-D-glucopyranosyl) disulfide with Et_3PAuR (R =
C_{1-6} alkylthio, $PhCH_2S$, $PhCH_2CH_2S$, or Et_3PAuS) in an inert aprotic
organic solvent, such as $CHCl_3$; or alternatively, by direct O-
acetylation of (triethylphosphine)gold 1-thio-β-D-glucopyranoside
with excess Ac_2O or ~4 mol equivalents of AcCl in the presence of
excess tertiary amine or pyridine. They can also be prepared by the
reaction of thioglucose derivatives of alkali metals, NH_4, Pb, Ag,
or Cu with $(Et_3P)_2AuX$ (X = halide, nitrate, or SCN) in $CHCl_3$ [67,68].
Alternate methods of synthesis involve the intermediate preparation
of the S-gold derivative of 2,3,4,6-tetra-O-acetyl-1-thio-β-D-gluco-
pyranose, which on treatment with Et_3P gives auranofin [69], and the
reaction of the thioglucose with Et_3PAuCl in the presence of an
amine [70].

   Biochemistry of Auranofin. Auranofin is reported to be orally
effective in animal (adjuvant rat) and human (rheumatoid) arthritic
conditions. It is a potent in vitro inhibitor of the release of
lysosomal enzymes from phagocytizing rat leukocytes. The dose-
related serum gold levels indicate that it is effectively orally
absorbed. Oral auranofin produces significant dose-response suppression of primary and secondary lesions of adjuvant arthritic rats, and
its antiarthritic potency is equal to or better than injectable
$Na_2Au(tm)$. That the antiarthritic action of auranofin is due to the
gold present in it is proved by the fact that the non-gold-containing
substructures of auranofin, e.g., Et_3PO and 2,3,4,6-tetra-O-acetyl-1-
thio-β-D-glucopyranose, administered to the adjuvant arthritic rats
at concentrations equal to those given with auranofin have no effect.
The advantages of oral auranofin over parenterally administered
$Na_2Au(tm)$ include lower gold levels in blood and kidneys and less
toxicity. While i.m. $Na_2Au(tm)$ is partially adrenal-dependent, oral
auranofin (20 mg Au/kg) produces a marked inhibition of the carrage-
enan-induced paw edema in both adrenalectomized and sham-operated
rats. The apparent superiority of auranofin as an anti-inflammatory
agent is observed at blood gold levels of approximate 6.0 µg/ml as

against 3.0 μg/ml for $Na_2Au(tm)$. Pharmacokinetic studies in normal
rats suggest that daily oral chrysotherapy produces a greater
stability in blood gold levels and less renal gold accumulations
than injectable gold drugs [71,72].

A number of in vivo and in vitro experimental models demonstrate the pharmacologic properties of auranofin. It binds to blood
cells in rats, dogs, and humans, whereas $Na_2Au(tm)$ is not associated
with the formed elements of the blood. It affects various humoral
and cellular immune lysosomal enzyme release from phagocytizing
(PMN) leukocytes and PMN-mediated antibody-dependent cellular cytotoxicity (ADCC) of target cells [73-77]. Auranofin decreased "7S"
anti-SRBC antibody generation in adjuvant arthritic rats, producing
a 60% reduction in the number of hemolytic plaque-forming cells
(PFCs) in normal rats. Both $Na_2Au(tm)$ and auranofin inhibit mitogen
responses ($[^3H]$thymidine uptake) of peripherial blood lymphocytes
obtained from rheumatoid arthritis patients [68], but auranofin has
a greater inhibitory effect (90 vs. 20%) on lymphocyte PHA. In
vitro studies [74] show that auranofin but not $Na_2Au(tm)$ inhibits
the incorporation of $[^3H]$thymidine and $[^{14}C]$amino acids (precursors
of DNA and protein syntheses) in mitogen- (PHA-) stimulated human
lymphocytes. Also, auranofin inhibits the membrane transport of
both $[^3H]$thymidine and $[^3H]$2-deoxy D-glucose and does not affect DNA
polymerase or the protein-synthesizing mechanism at the intracellular
level. Thus, the inhibitory action of auranofin on lymphocyte
responses may be due to its effect at the cellular membrane level.
In contrast to these in vitro studies, the in vivo experiments
strongly favor an immunoregulation of the cell-mediated immune
responses rather than suppression.

Like the antiarthritic drug levamisole (vide supra), auranofin
is a condition-dependent immunoregulant inhibiting abnormally high
responses and stimulating abnormally low responses. At in vitro
concentrations of 10^{-3} to 10^{-6} M (10 μM = 2 μg Au/ml), auranofin
produces a marked reduction in the extracellular levels of lysosomal
enzyme markers (ß-glucuronidase and lysozyme) released from zymosan-
stimulated rat leukocytes [76], and at a concentration of 1 μg Au/ml

(5 μM) marked reduction (≥72%) in β-glucuronidase, acid phosphatase, and lysozyme release results. In contrast, $Na_2Au(tm)$ had either no or very little inhibitory activity on lysosomal enzyme release [77]. Blood gold levels in auranofin-treated rheumatoid arthritic patients were within the range required for in vitro inhibition of lysosomal enzyme release and correlated with decreases in human gamma globulin (IgG), rheumatoid factor (RF) titers, and IgG-RF immune complex formation in vitro.

These results suggest that the inhibition of lysosomal enzyme release (and/or decrease in immune complex formation) may be responsible for the therapeutic action of auranofin. In vitro tests also show that auranofin but not $Na_2Au(tm)$ inhibits PMN-mediated antibody (heat-inactivated immune rat sera)-dependent lysis of ^{51}Cr-mouse to cells. A significant difference between $Na_2Au(tm)$ and auranofin is that $Na_2Au(tm)$ does not potentially inhibit the sulfhydryl group reactivity [72]. Lastly, the predicted value of auranofin in rheumatoid arthritis based on the arthritic and immunological parameters in the in vitro and in vivo models is borne out in human studies [78,79].

2.2.4. The Bis-Coordinated Gold(I) Salts as Attempted Oral Drugs

Among examples of some other orally administered gold drugs are the bis-coordinated gold(I) salts of the type $[R_3PAuPR_3']^+X^-$ (R, R' = C_{1-4} alkyl or alkoxy, Ph; X = halide, ClO_4, BF_4 [80]), $(R_3PAu)_2S$ [81], $[R^1R^2SAuSR^1R^2]^+X^-$ (R^1, R^2 = $HOCH_2CH_2$; or R^1R^2S = tetrahydrothienyl or tetrahydrothiopyranyl), and the bis(pyridine) gold(I) salts (e.g., XIV) [80]. Reactions of Et_3PAuCl and pyridine in AcMe in the pres-

(XIV)

ence of $AgClO_4$ yield the compound $[Et_3PAu(pyridine)]ClO_4$ [81]. Similarly prepared were also other compounds (see Structure XV, where

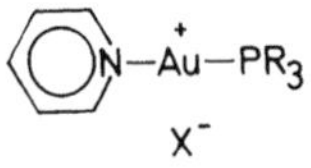

(XV)

R = C_{1-4} alkyl or alkoxy, Ph, halophenyl, C_{1-4} alkoxy phenyl; X =
ClO_4, IO_4, BF_4, and PF_6) [81].

Other oral drugs prepared include compositions comprising S-
(phosphino or phosphito aurous)thiouronium halides [$R_3PAuSC(:N^+HR^2)$
$NHR^1]X^-$ (R = C_{1-3} alkyl or alkoxy, Ph or PhO; R' = H, NH_2; R^2 = H,
Me; X = Cl, Br, I), obtained by reactions of R_3PAuCl with $(H_2N)_2CS$
in EtOH [82]. They can be administered orally in daily doses of 1-25
mg (calculated on Au content).

A series of ionic phosphorous ylide complexes of gold(I), of
the type [$R_3P-Au-CH_2PR_3]^+Cl^-$ (R = C_2H_5) and [$(R_3PCH_2)_2Au^+]X^-$ (R =
CH_3, C_2H_5, n-C_4H_9, C_6H_5; X = Cl, Br) and the cyclic complex (XVI)

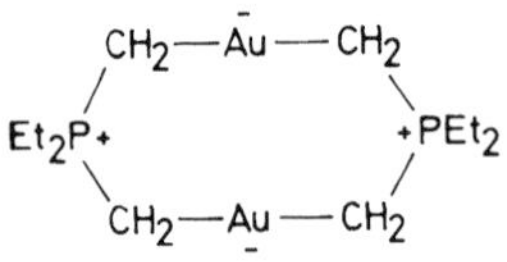

(XVI)

were recently synthesized and tested as oral chrysotherapeutic agents
[70]. Of all these complexes, only the [$Et_3PCH_2AuCH_2PEt_3]^+Cl^-$ was
found to be comparable to standard chrysotherapeutic agents, although
LD_{50} tests for mice showed too narrow a gap between its toxic and
therapeutically useful doses.

3. GOLD POISONING, CONTRAINDICATIONS, AND TREATMENT

Most of the gold drugs in use are Au(I) thiolates, and the results
of pharmacological as well as toxicological studies preclude the use
of gold(III) compounds as drugs. Gold(III) compounds are highly

toxic because of their strong oxidizing power, greater inhibition of enzymes, and greater retention in tissues, and were in no case acceptable as drugs.

It is established that an effective treatment of rheumatoid arthritis with gold drugs results in some degree of clinical toxicity whose severity and extent varies depending upon the mode of administration, rate of elimination of gold from the body, and other factors. The common toxic reactions involve the skin and mucous membrane. Cutaneous reactions may vary in severity from simple erythema to severe exfoliative dermatitis. Lesions of the mucous membranes may include stomatitis, pharyngitis, tracheitis, gastritis, glossitis, colitis, and vaginitis. Gold drugs may result in blood dyscrasias, thrombocytopenia, leukopenia, agranulocytosis, eosinophilia, and a number of other reactions, including encephalitis, peripherial neuritis, hepatitis, and nitritoid crisis. Gold is toxic to kidney, and traces of albumin are observed in urine during the chrysotherapy. However, heavy albuminuria and microscopic hematuria occur very rarely. In few cases gold nephrosis may occur which can be prevented by cortisol. Toxic reactions involving the liver and bone marrow are also observed. In Table 1 a list of typical predominant side effects in chrysotherapy with gold thiopolypeptide is given.

TABLE 1

Some Typical Side Effects in Chrysotherapy with
Gold Thiopolypeptide of 74 Patients

Kind of side effect	Abundance
Dermatitis	9
Eosinophilia	8
Proteinuria	8
Stomatitis	7
Pruritis	7
Gastroenteritis	2
Erythrozyturia	1
Leukopenia	1

Source: Collected from the references given in Ref. 42.

Because of their side effects, gold drugs are not tolerated by aged individuals and by individuals having renal disease, infectious hepatitis, blood dyscrasias, urtricaria, eczema, and colitis, all considered to be contraindications to the use of the gold drugs.

To avoid serious toxic reactions, most physicians initiate chrysotherapy with small doses of gold and then increase the dose gradually. The therapy is withheld temporarily in case of a serious untoward response. When the toxic reactions are too severe, detoxifying agents (or antagonists) such as penicillamine (β,β-dimethyl cysteine), dimercaprol (2,3-dimercaptopropanol, known as British antilewisite or BAL--Structure XVII), antihistamines, or adrenocorticosteroids may be administered topically [83,84].

$$
\begin{array}{ccc}
\text{H} & \text{H} & \text{H} \\
| & | & | \\
\text{H}-\text{C}-\text{C}-\text{C}-\text{H} \\
| & | & | \\
\text{SH} & \text{SH} & \text{OH}
\end{array}
$$

(XVII)

As the toxicity of gold arises due to its combination with one or more reactive groups of the O, S, and N ligands, which in the body take the form of -OH, -COO^{-}, $PO_4H_2^{-}$, $>$C=O, -SH, -S-S-, -NH$_2$, =NH, etc. and are essential for normal physiological functions, an antagonist against gold toxicity should have the property of reacting with it to form tightly bound complexes, thereby preventing or reversing the binding of the toxic gold to body ligands. The administration of the antagonist promotes the excretion of gold through urine or feces. Using rats as model animals, the relative effectiveness of the detoxifying agents in promoting urinary excretion appears to be penicillamine > BAL > thiomalate.

4. GOLD METABOLISM (ABSORPTION, DISTRIBUTION, AND EXCRETION)

This aspect has been reviewed elegantly by Shaw [7] recently, and we will only consider some salient and relevant features here. It is demonstrated [85] that the water-soluble gold drugs, e.g., $Na_2Au(tm)$, $Na_3[Au(S_2O_3)_2]$, etc., are rapidly absorbed after i.m. or i.v. injection and plasma concentrations increase rapidly within a few hours, after which they are maintained at a slightly lower but constant level. However, oil suspensions of these drugs administered intramuscularly result in minimal blood level.

Experiments concerning the distribution of gold in animals and also in humans show that gold in the bloodstream is almost entirely bonded to the protein in the plasma, i.e., the serumalbumin, but not to the fibrinogen [42]. It is also not bonded to the erythrocytes.

Gold accumulates in the liver, spleen, and kidneys of rats [53,86], rabbits [87], and humans [88]. In rabbit kidneys the gold is 4 times more concentrated in the cortex than in the medulla, which is consistent with electron microscopic observation of gold deposits in glomeruli and proximal tubules [89-91]. Systematic studies of the subcellular distribution have not been reported, although gold has been reported in the mitochondria [89,90] and lysosomes [91,92] of hepatic and renal tissue. Gold is also present in the cytoplasm of mouse liver cells [93,94]. The painful joints of an arthritic patient contain more than twice as much gold as uninvolved joints [95].

The water-soluble gold drugs are eliminated mainly via urinary excretion. Insoluble gold drugs, such as colloidal gold and gold sulfide, accumulate in the same organs, but with greatest concentration in the liver and with the fecal excretion as primary route of gold elimination.

5. PHYSICAL METHODS FOR DETERMINING INTERACTIONS
OF GOLD DRUGS IN BIOLOGICAL SYSTEMS

A number of powerful physical methods are available for the characterization of gold in its great variety of compounds, none of which
is sensitive enough, however, for the study of gold at physiological
conditions in vivo. In vitro experiments, on the other hand, have
quite often been followed by the prevailing modern techniques of
Mössbauer spectroscopy, electron spectroscopy for chemical analysis,
and nuclear quadrupole resonance spectroscopy.

Monoisotopic [197]Au gives rise to one of the best Mössbauer
resonances, and hence there is a whole range of chemical applications
of [197]Au Mössbauer spectroscopy [96]. It is now well established
that Mössbauer spectra can distinguish valence states and coordination numbers in gold complexes [97-99]. Even unusual intermediate
oxidation states are easily characterized through the isomeric
shifts and quadrupole splittings, and information on coordination
geometry may be obtained similarly [100-104]. The method has the
disadvantage of being dependent on rather large samples, and of
course it is applicable only in the solid state.

[197]Au has a large quadrupole moment ($I = 3/2$), and should exhibit a signal in nuclear quadrupole resonance (NQR) spectroscopy of
gold compounds. Surprisingly, this has not been observed, and the
few references in the literature are more the exception to the rule
and may even be erroneous [101,105]. Again, NQR is limited to the
solid state.

Electron spectroscopy for chemical analysis [ESCA] is another
method for the determination of the state of bonding of gold centers,
particularly for oxidation states of gold in solids. The Au (4f 7/2)
binding energies increase with the oxidation state of the metal and
thus offer a reliable source of information [100,106-109]. Only
small samples are required, and ESCA could therefore qualify for
studies of biological systems in the future.

The classical methods of ultraviolet (UV), visible, and infrared (IR) spectroscopy have been widely and successfully employed for

the characterization of gold compounds, but they are to be taken as
a probe more of the ligand systems than of the metal atom. There
are very few typical, strongly metal-dependent IR absorptions of
gold complexes in the conventional region due to the heavy mass of
the element, and UV bands are often broad and overlapping. The inclusion of circular dichroism (CD) strengthens the interpretation of
electronic spectra, since overlapping bonds may have different in-
tensities and signs in CD spectra [110]. The greatest advantage of
the spectral methods is their applicability to the solution state,
most important for biological systems.

With very few exceptions, gold compounds are diamagnetic.
Though no regular ^{197}Au nuclear magnetic resonance (NMR) is possible,
NMR of other nuclei is almost always available. ^{1}H, ^{13}C, and ^{31}P
NMR spectroscopy are finding more and more applications for the
study of biochemical and pharmacological problems, since the advent
of the Fourier transform technique has greatly enhanced the sensitivity and resolution of this technique [44,111,112]. Again, the
solution state is the normal and easiest system of study for NMR,
but investigations of glasses and amorphous or oriented solids are
also possible. A recent work on the nature of interactions between
Et$_3$PAuCl and the various biological substrates by ^{31}P NMR is an
example of the potential of NMR as a noninvasive method for pertin-
ent systems [112].

6. DETECTION AND ESTIMATION OF GOLD IN BIOLOGICAL SYSTEMS

Methods for the determination of gold in biological systems currently in use are neutron activation analysis [113-115], atomic absorption spectroscopy [116-118] and, if applicable, electron microscopy [119-123]. Radioactive tracer studies can be carried out employing the radioactive synthetic gold isotopes ^{198}Au and ^{195}Au,
e.g., in whole-body radioactivity monitoring (whole-body counter)
[42]. The high cost of neutron activation analysis prevents it from

being used in clinical examination on a greater scale. Determination of gold in plasma and urine during chrysotherapy, however, can be performed very simply, precisely, and quickly by atomic absorption spectroscopy (AAS). Both flame and carbon furnace excitation are in use [116-118]. Very recently plasma excitation (ICP) spectroscopy has proven to be superior to AAS for many routine analytical purposes. It is thus very likely that ICP will replace AAS as soon as the equipment--costly at present--becomes more readily available at lower price.

In early electron microscopy investigations gold deposits could not be identified exactly and were referred to as "extra-electron-dense" regions in biological substrates. In more recent studies [119-123] electron microscopy was coupled with x-ray absorption methods to analyze for gold directly, and the results seem to be promising.

REFERENCES

1. R. Koch, *Deutsch Med. Wochenschr.*, *16*, 756 (1927).

2. A. Feldt, *Ber. Klin. Wochenschr.*, *54*, 1111 (1917).

3. K. Landé, *Münchener Med. Wochenschr.*, *74*, 1132 (1927).

4. J. Forestier, *Bull. Mém. Soc. Méd. Hôp* (Paris), *53*, 323 (1929).

5. H. Möllgaard, *Chemotherapy of Tuberculosis*, Nordisk, Copenhagen, 1924.

6. P. J. Sadler, *Struct. Bonding*, *29*, 171 (1976); *Gold Bull.*, *5*, 110 (1976).

7. C. F. Shaw III, *Inorg. Persp. Biol. Med.*, *2*, 287 (1979).

8. D. H. Brown and W. E. Smith, *Quart. Rev.*, 217 (1980).

9. J. A. Cruciani, *Sem. Med.* (Buenos Aires), *II*, 854 (1938).

10. I. Viginati, V. Skalak, and S. Rachenberg, *Press. Med.*, *46*, 1482 (1938).

11. H. Friebel, *Med. Exp.*, *4*, 37 (1961).

12. M. Muranaka, T. Miyamato, and T. Shida, *Ann. Allerg.*, *40*, 132 (1978).

13. B. Oddo, *Arch. Farmacol. Sper.*, *44*, 141 (1927).

14. P. Pichat, *Compt. Rend. Soc. Biol., 132,* 13 (1939).

15. N. S. Pennys, W. E. Eaglestein, and P. Frost, *Arch. Dermatol., 108,* 56 (1973); *112,* 185, 1467 (1976).

16. S. Yamashita, Y. Seyema, and I. Nishikawa, *Experientia, 34,* 472 (1978); S. Yamashita, *Jap. Kokai, Tokkyo Koho, 79,* 20, 121 (1979).

17. U. P. Basu and I. Sikdar, *J. Indian Chem. Soc., 24,* 466 (1947).

18. S. Melhan, U.S. Patent 2, 507; *Chem. Abstr., 44,* 7030 (1950).

19. D. G. Craciunescu, *Ann. Roy. Acad. Farm., 43,* 265 (1977).

20. U. Langness and W. Decius, *Arzneim.-Forsch., 28,* 2202 (1978).

21. R. L. Kieft, W. M. Peterson, G. L. Blundell, S. Horton, R. A. Henry, and H. B. Jonassen, *Inorg. Chem., 15,* 1721 (1976).

22. C. Dragulescu, J. Heller, A. Maurer, S. Policec, V. Topcui, M. Csalci, S. Kirschner, S. Kravitz, and R. Moraski, 16th Int. Coord. Chem. Conf., 1.9 (1974).

23. T. M. Simon, D. H. Kunishima, G. J. Vibert, and A. Lorber, *Cancer* (Philadelphia), *44,* 1965 (1979).

24. A. F. Alfonso, A. Hassan, B. Gardner, S. Stein, J. Patti, N. A. Solanon, J. McCarthy, and J. Steigmann, *Cancer Res., 38,* 2740 (1978).

25. D. T. Walz, M. J. DiMartino, A. Misher, and B. M. Sutton, *Proc. Soc. Exp. Biol. Med., 140,* 263 (1972).

26. D. T. Walz and M. J. DiMartino, *J. Med. Chem., 15,* 1095 (1972).

27. Empire Rheumatism Council Research Sub-Committee, *Ann. Rheum. Dis., 20,* 315 (1961).

28. American Rheumatism Association Cooperating Clinics Committee, *Arthr. Rheum., 16,* 353 (1973).

29. J. W. Sigler, B. G. Bluhm, H. Duncan, I. T. Sharp, D. C. Ensign, and W. R. McCrum, *Arthr. Rheum., 15,* 125 (1972).

30. P. Smit, D. T. Walz, M. J. DiMartino, and B. M. Sutton, *Med. Chem. Ser. Monogr., 13,* 209 (1974).

31. R. H. Girdwood, *Brit. Med. J., 1,* 501 (1974).

32. G. Orestano, *Boll. Soc. Ital. Biol. Sper., 7,* 256, 748, 1281, 1284, 1286, 1289 (1932).

33. G. Orestano, *Arch. Int. Pharm., 44,* 259 (1933).

34. A. B. Sabin and J. Warren, *J. Bacteriol., 40,* 823 (1940); *Science, 92,* 535 (1940).

35. E. E. Moore and R. J. Ohman, U.S. Patent 2,509,200 (1950).

36. R. H. Freyberg, W. D. Block, and S. Levey, *J. Clin. Invest., 20,* 401 (1941).

37. W. D. Block, O. H. Buchanan, and R. H. Freyberg, *J. Pharmacol.*, *73*, 200 (1941); *76*, 355 (1942); *82*, 391 (1944).

38. W. D. Block and E. L. Knapp, *J. Pharmacol.*, *83*, 275 (1945).

39. H. Brown, *J. Amer. Chem. Soc.*, *49*, 958 (1927).

40. H. Ruben, A. Zalkin, M. O. Faltens, and D. H. Templeton, *Inorg. Chem.*, *13*, 1836 (1974).

41. R. F. Baggio and S. Baggio, *J. Inorg. Nucl. Chem.*, *35*, 3191 (1973).

42. M. Schattenkirchner, *Comp. Rheum.*, Eular, Basel, 1977; *Schweiz. Med. Wschr.*, *107*, 1145 (1977); W. Meyer and W. Weyerbrock, *Internist, 20*, 426 (1979); R. Eberl, *Wien. Klin. Wochenschr.*, *86*, 3 (1974).

43. L. F. Larkworthy and D. Sattari, *J. Inorg. Nucl. Chem.*, *42*, 551 (1980).

44. A. A. Isab and P. J. Sadler, *J.C.S. Chem. Commun.*, 1051 (1976).

45. D. T. Walz, M. J. DiMartino, and A. Misher, *Ann. Rheum. Dis.*, *30*, 303 (1971).

46. R. D. Sofia and J. F. Douglas, *Agents Actions, 3*, 335 (1973).

47. P. K. Fox, A. J. Lewis, P. McKeown, and D. D. White, *Brit. J. Pharmacol.*, *66*, 141P (1979).

48. M. Adam, *Z. Rheumaforschg.*, *27*, 102 (1968).

49. J. J. Lennings, S. Macrae, and R. M. Gorczynskii, *Clin. Exp. Immunol.*, *36*, 260 (1979).

50. R. N. P. Caphill, *Experientia, 27*, 913 (1971).

51. G. Weissmann, *Lancet, ii*, 1373 (1964).

52. R. S. Ennis, J. L. Granda, and A. S. Posner, *Arthr. Rheum.*, *11*, 756 (1968).

53. L. Vegh and E. Hardegger, *Helv. Chim. Acta, 56*, 2079 (1973).

54. G. Brecher and S. H. Waxler, *Amer. J. Physiol.*, *162*, 428 (1950).

55. P. D. Rogers, P. G. Webb, and S. A. Jagot, *IRCS Med. Sci. Libr. Compend.*, *7*, 402 (1979).

56. N. B. Marshall, R. J. Barnett, and J. Mayer, *Proc. Soc. Exp. Biol. Med.*, *90*, 240 (1955).

57. D. T. Walz, M. J. DiMartino, and A. Misher, *J. Pharmacol. Exp. Ther.*, *178*, 223 (1971).

58. D. T. Walz, M. J. DiMartino, B. M. Sutton, and A. Misher, *J. Pharmacol. Exp. Ther.*, *181*, 292 (1972).

59. B. M. Sutton, E. McGusty, D. T. Walz, and M. J. DiMartino, *J. Med. Chem.*, *15*, 1095 (1972).

60. H. Kamel, D. H. Brown, J. M. Ottoway, W. E. Smith, J. Cottney, and A. J. Lewis, *Agents Actions, 8*, 546 (1978).

61. E. R. McGusty, B. M. Sutton, and D. T. Walz, German Patent
    2,061,181 (1971); *Chem. Abstr., 75,* 52818 (1971).

62. B. M. Sutton, D. T. Walz, and J. Weinstock, German Patent
    2,434,920 (1975); *Chem. Abstr., 83,* 84857 (1975).

63. E. R. McGusty and B. M. Sutton, U.S. Patents 3,718,679,
    3,718,680 (1973); *Chem. Abstr., 78,* 135673, 135672 (1973).

64. B. M. Sutton and J. Weinstock, German Patent 2,437,147 (1975)--
    Chem. Abstr., 83, 10389 (1975); U. S. Patent 3,903,274 (1976)--
    Chem. Abstr., 84, 17559 (1976).

65. J. Weinstock, B. M. Sutton, G. Y. Kuo, D. T. Walz, and M. J.
    DiMartino, *J. Med. Chem., 17,* 139 (1974).

66. W. S. Crane and H. Beall, *Inorg. Chim. Acta, 31,* 469 (1978).

67. D. T. Hill, I. Lantos, and B. M. Sutton, U.S. Patents 4,115,642
    (1978), 4, 122,254 (1978), 4,124,759 (1978), 4,125,710 (1978),
    4,125,711 (1978); *Chem. Abstr., 90,* 72435b, 87834e, 104297b,
    121969a, 121970h (1979).

68. Smithkline Corporation, *Jap. Kokai Tokkyo Koho, 78,* 132,528;
    Chem. Abstr., 90, 168922 (1979).

69. D. T. Hill, I. Lantos, and B. M. Sutton, U.S. Patent 4,133,952
    (1979); *Chem. Abstr., 90,* 127542s (1979).

70. H. Schmidbaur, J. R. Mandl, and A. Fügner, *Z. Naturforsch., 33b,*
    1325 (1978).

71. D. T. Walz, M. J. DiMartino, D. Griswold, S. Alessi, and E.
    Bumbier, Abstr. 6th Int. Cong. Pharmacol., Helsinki, No. 1324,
    p. 552 (1975).

72. D. T. Walz, M. J. DiMartino, L. W. Chakrin, B. M. Sutton, and
    A. Misher, *J. Pharmacol. Exp. Ther., 197,* 142 (1976).

73. A. Lorber, W. H. Jackson, and T. M. Simon, 6th Western Reg.
    Meeting, Scottsdale, Ariz., Nov. 1977; as quoted in Ref. 75.

74. A. E. Finkelstein, O. R. Burrone, D. T. Walz, and A. Misher,
    J. Rheumatol., 4, 245 (1977).

75. D. T. Walz and D. E. Griswold, *Inflammation, 3,* 117 (1978).

76. M. J. DiMartino and D. T. Walz, *Inflammation, 2,* 131 (1977); *4,*
    279 (1980).

77. A. E. Finkelstein, F. R. Roisman, and D. T. Walz, *Inflammation,*
    2, 143 (1977).

78. A. E. Finkelstein, D. T. Walz, V. Batista, M. Mizraji, F. Rois-
    man, and A. Misher, *Ann. Rheum. Dis., 35,* 251 (1976).

79. F.-E. Berglöf, K. Berglöf, and D. T. Walz, *J. Rheumatol., 5,*
    68 (1978).

80. D. T. Hill, U.S. Patents 4,057,630 (1978), 4,093,719 (1978),
    4,112,113 (1979), 4,098,887 (1979); *Chem. Abstr., 88,* 94838
    (1978), *89,* 152729 (1978), *90,* 127542 (1979), *90,* 123057 (1979).

81. D. T. Hill, European Patent Application 3,897 (1979); *Chem. Abstr., 92,* 94575 (1980).

82. B. M. Sutton and J. Weinstock, U.S. Patent 3,787,568 (1974).

83. D. T. Hill, *Med. Clin. N. Amer., 52,* 733 (1968).

84. R. H. S. Thompson and V. P. Whittaker, *Biochem. J., 41,* 342 (1947).

85. R. H. Freyberg, in *Arthritis and Related Conditions,* 7th ed. (J. L. Hollander, ed.), Lea and Febiger, Philadelphia, 1966, pp. 302-332.

86. W. D. Block, O. H. Buchanan, and R. H. Freyberg, *J. Pharm. Exp. Ther., 74,* 355 (1942).

87. E. G. McQueen and P. W. Dykes, *Ann. Rheum. Dis., 28,* 437 (1969).

88. R. H. Freyberg, W. D. Block, and S. Levey, *Ann. Rheum. Dis., 1,* 77 (1942).

89. A. H. Nagi, F. Alexander, and A. Z. Barbes, *Exp. Molec. Pathol., 15,* 354 (1971).

90. P. Galle, *J. Microscopie, 19,* 17 (1974).

91. S. W. Strunk and M. Ziff, *Arthr. Rheum., 13,* 39 (1970).

92. M. Davies, J. B. Lloyd, and F. Beck, *Biochem. J., 121,* 21 (1971).

93. R. M. Turkall, J. R. Biachine, and A. P. Lerber, *Fed. Proc., 36,* 356 (1977).

94. E. M. Mogilinicka and J. K. Piotroski, *Biochem. Pharmacol., 26,* 1819 (1977).

95. J. S. Lawrence, *Ann. Rheum. Dis., 20,* 341 (1961).

96. M. O. Faltens and D. A. Shirley, *J. Chem. Phys., 53,* 4249 (1970).

97. R. Hüttel and H. Forkl, *Chem. Ber., 105,* 1664 (1972).

98. R. V. Parish and J. D. Rush, *Chem. Phys. Lett., 63,* 37 (1979).

99. J. A. J. Jaruis, A. Johnson, and R. J. Puddephatt, *J.C.S. Chem. Commun.,* 373 (1975).

100. H. Schmidbaur, J. R. Mandl, F. E. Wagner, D. F. van de Vondel, and G. P. van der Kelen, *J.C.S. Chem. Commun.,* 170 (1976).

101. P. Machmer, M. Read, and P. Cornil, *C. R. Acad. Sci., Ser. A.B., 262B,* 650 (1966); *Inorg. Nucl. Chem. Lett., 3,* 215 (1967).

102. C. A. McAuliffe, R. V. Parish, and P. D. Randall, *J. Chem. Soc., Dalton Trans.,* 1426 (1977).

103. P. G. Jones, A. G. Maddock, M. J. Mays, M. M. Muir, and A. F. Williams, *J. Chem. Soc., Dalton Trans.,* 1434 (1977).

104. G. C. H. Jones, P. G. Jones, A. G. Maddock, M. J. Mays, P. A. Vergnan, and A. F. Williams, *J. Chem. Soc., Dalton Trans.*, 1440 (1977).

105. P. Machmer, *Z. Naturforsch.*, *21*, 1025 (1966); *J. Inorg. Nucl. Chem.*, *30*, 2627 (1968).

106. C. Battistoni, G. Mattogno, F. Cariati, L. Naldini, and A. Sgamellotti, *Inorg. Chim. Acta*, *24*, 207 (1977).

107. D. F. van de Vondel, G. P. van der Kelen, H. Schmidbaur, A. Wohlleben, and F. E. Wagner, *Physica Scripta*, *16*, 364 (1977).

108. A. McNeillie, D. H. Brown, W. E. Smith, M. Gibson, and L. Watson, *J. Chem. Soc., Dalton Trans.*, 767 (1980).

109. P. M. T. M. van Attekum and J. M. Trooster, *J. Chem. Soc., Dalton Trans.*, 201 (1980).

110. D. H. Brown, G. C. McKinlay, and W. E. Smith, *J. Chem. Soc., Dalton Trans.*, 1874 (1977); *Inorg. Chim. Acta*, *32*, 117 (1979).

111. N. A. Malik and P. J. Sadler, *Biochem. Soc. Trans.*, *7*, 731 (1979).

112. N. A. Malik, G. Otiko, and P. J. Sadler, *J. Inorg. Biochem.*, *12*, 317 (1980).

113. S. Soelvsten, *Scand. J. Chim. Lab. Invest.*, *16*, 39 (1964).

114. D. Brune, K. Samsahl, and P. O. Wester, *Clin. Chim. Acta*, *13*, 285 (1966).

115. F. E. Kriesins, A. Markkanen, and P. Pelota, *Ann. Rheum. Dis.*, *29*, 232 (1970).

116. M. Schattenkirchner and Z. Grobenski, *Atom. Abs. News Lett.*, *16*, 84 (1977).

117. H. Kamel, D. H. Brown, J. M. Ottoway, and W. E. Smith, *Talanta*, *24*, 309 (1977).

118. A. Kamel, D. H. Brown, J. M. Ottoway, and W. E. Smith, *Analyst*, *101*, 790 and references therein (1976).

119. W. L. Norton and M. Ziff, *Arthr. Rheum.*, *9*, 589 (1966).

120. P. E. Lipsky, K. Ugai, and M. Ziff, *J. Rheumatol.*, *6*, Suppl. 5, 131 (1979).

121. F. N. Ghandially, *J. Rheumatol.*, *6*, Suppl. 5, 45 (1979).

122. F. N. Ghandially, W. E. DeCoteau, S. Huong, and I. Thomas, *J. Path.*, *124*, 77 (1979); *125*, 219 (1979).

123. F. N. Ghandially, A. F. Ortschak, and D. M. Mitchell, *Ann. Rheum. Dis.*, *35*, 67 (1976).

Chapter 7

METAL IONS AND CHELATING AGENTS IN
ANTIVIRAL CHEMOTHERAPY

D. D. Perrin and Hans Stünzi
Medical Chemistry Group
The John Curtin School of Medical Research
The Australian National University
Canberra, Australia

1. INTRODUCTION

When a host organism is exposed to attack by a bacterial, fungal, or
protozoal infection it is sometimes possible to apply treatment which
is not injurious to the host but which destroys or controls the invading agent. This concept of selective toxicity [1] has proved of

great service in the rational choice of antifungal and antibaterial
agents and in the interpretation of their actions. Its successful
application to antiviral chemotherapy has been much more limited,
due, in large measure, to the much closer relationship that exists
between host organisms and invading virus.

The existence of virus-specific metalloenzymes and the differ-
ential toxicity of free metal ions for viral proteins continue to
hold out some promise that metal ions and substances that chelate
them might have useful applications as antiviral agents. 1-Methyli-
satin β-thiosemicarbazone was the first synthetic antiviral drug
used in medicine, for the prophylaxis of smallpox [2]. It is a
chelating agent, but since its introduction 25 years ago, discussions
on a connection between its antiviral activity and chelating ability
have been largely speculative. This is due to the complexity of
living matter. Cell fluids are rich in complexing agents which compete for small amounts of metal ions, so that it may be difficult to
extrapolate from test tube chemistry and biology to in vivo conditions. As a suitable starting point, we compare the stability of
drug-metal complexes with those of metal complexes of naturally
occurring ligands.

On the other hand, it is necessary to relate inorganic results
with the molecular biology of viruses:

Viruses lie at the border between living and nonliving matter.
They cannot reproduce outside the host cell. Viruses contain either
DNA or RNA but not both. This is surrounded by nucleoproteins and a
protective shell of capsid proteins, which, in turn, may lie inside
a lipoprotein envelope. Thus, the simple papovaviruses contain a
core of DNA (~4 million daltons) and the capsid coat. The DNA carries the genetic information for the biosynthesis of the six types
of viral proteins. The host cell provides the starting materials,
the enzymes, and the energy for the biosynthesis of the proteins and
the replication of the DNA. More complicated viruses contain many
more genes, including some that code for enzymes that are biologically related to the host's enzymes, and a virion may contain a
limited number of enzymes. It is important to remember that a virus

has no mitochondria and no ribosomes and hence is completely dependent on the host cell for its energy and synthetic requirements.

It is convenient to classify viruses according to the type of nucleic acid they contain. This is done in Table 1 for some of the better-known viruses.

TABLE 1

Some Human and Animal Viruses, Classified by
Type of Nucleic Acid

Type of virus	Major groups	Type of nucleic acid[a]	Examples
DNA viruses	Adenoviridae	ds	
	Herpetoviridae	ds	Cytomegalo virus Epstein-Barr virus Herpes simplex virus Marek's disease virus Pseudorabies virus *Varicella zoster* virus (chicken pox)
	Papovaviridae	ds	Papilloma viruses
	Poxviridae	ds	Vaccinia (cowpox) virus Variola (smallpox) virus Myxoma virus
	Others		Hepatitis virus A and B
RNA viruses	Myxoviridae	$^-$ss	Influenza virus Measles virus Mumps virus Parainfluenza virus
	Picornaviridae	$^+$ss	Enteroviruses: Coxsackie virus echo virus Mengo virus polio virus Rhinoviruses: human rhinovirus foot-and-mouth disease virus
	Reoviridae	ds	

TABLE 1 (Continued)

Type of virus	Major groups	Type of nucleic acid[a]	Examples
RNA viruses	Retroviridae	[+]ss	Oncovirus, e.g., Rous sarcoma virus, "slow" viruses (maedi, kuru, visna)
	Rhabdoviridae	[-]ss	Rabies virus Vesicular stomatitis virus
	Togaviridae	[+]ss	Rubella virus (German measles) Sindbis virus Tick-borne encephalitis virus Yellow fever virus

[a]ds = double-stranded; [-]ss = single-stranded, antimessage (negative) polarity; [+]ss = single-stranded, message (positive) polarity.

Source: Compiled from data given in Ref. 3.

An understanding of the molecular biology of viruses may suggest ways by which to control or prevent viral infection.

2. THE VIRAL CYCLE AND ITS IMPACT ON HOST ORGANISMS

Viral diseases differ in two fundamental respects from bacterial and fungal diseases. The virus particle (virion) is entirely dependent on the host cells' biochemical machinery, and the genome (genetic material) of the virus can impress its control on the host cell so as to divert the metabolism of the latter wholly or in part to the production of viral nucleic acid and viral protein.

Replication of viruses follows a cycle, outlined in Fig. 1, in which a virion outside a cell becomes attached to its surface and penetrates the cell. This may be by fusion of the viral lipid envelope with the cytoplasmic membrane or by the cell engulfing the

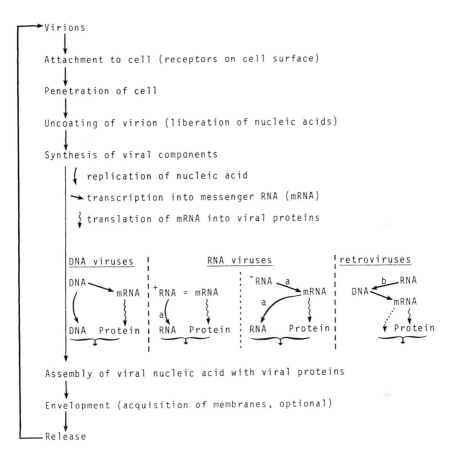

a) RNA-dependent RNA polymerases

b) RNA-dependent DNA polymerase (reverse transcriptase)

FIG. 1. The viral cycle. (Data from Ref. 3.)

virion. The protein coat of the virion is removed by lipases and peptidases and the viral nucleic acid is transported to its replication site where it takes over control of nucleic acid and protein synthesis. A mRNA copy of the viral genome is synthesized (*transcription*) from which the viral proteins are synthesized on the cellular ribosomes (*translation*). The viral nucleic acid is *replicated*. New viral proteins and nucleic acids are assembled into virions. The

cell finally ruptures, liberating large amounts of virion so that
the disease is spread. Specific details vary with the particular
type of virus studied, but some general points are worth mentioning:

Free virions may have surfaces which facilitate specific
attachment to the exterior of the host cell. For example, the influenza virion has two sets of glycoprotein spikes, made up of hemagglutinin and neuraminidase. The hemagglutinin causes the virion to
adsorb onto mucoprotein receptors on cells of the respiratory tract
by binding to sialic acid residues on the plasma membranes, leading
to entry of the virion into the cell. It is not at present known
whether the neuraminidase facilitates passage through the mucus
which lines the respiratory tract by hydrolyzing the neuraminic acid
present in the glycoprotein of mucus or whether it helps ensure the
release of newly formed virions from cells.

After the attachment, some bacterial viruses (bacteriophages)
have a sophisticated mechanism to thrust their nucleic acid into the
cell, whereas animal viruses probably penetrate intact or undergo
only partial degradation. Inside the cell, cytoplasmic proteases
and lipases complete the stripping of the external layers from the
viral nucleic acid which is then transported to the site of replication. This is the cytoplasm for pox viruses and myxoviruses, whereas
for herpes viruses it is the nucleus. Various pathways for the production of new viral proteins and nucleic acids are shown in Fig. 1.
The details vary, but the commonest routes are as follows:

Double-stranded DNA viruses. The nucleus and enzymes of the
host cell transcribe the DNA into mRNA. (Pox viruses are exceptions:
they have their own transcriptase which acts in the cytoplasm.) The
mRNA is translated into virus-coded enzymes and other "early" proteins on cellular ribosomes. Viral DNA is replicated. "Late" mRNA
is transcribed from progeny and parental DNA, and is translated
mainly into structural proteins.

Single-stranded RNA viruses. When single-stranded viral RNA
is of the "message" polarity (positive strand) it can serve both as
mRNA and as a template for its own replication. Retroviruses (e.g.,
Rous sarcoma virus) contain single-stranded RNA which is transcribed

by a viral enzyme (RNA-dependent DNA polymerase, reverse transcriptase) into an RNA-DNA hybrid. This serves as a template for syn-
thesizing double-stranded DNA which is integrated into the cellular
DNA prior to multiplication.

Influenza virus also contains single-stranded RNA, but this
has the "antimessage" polarity (negative strand). The original RNA
serves as a template for the transcription of complementary RNA by
virus-specific RNA-dependent RNA polymerases. The complementary RNA
is mRNA, which is copied again into the negative-strand viral genome
and also translated into viral proteins.

It is important to note that the RNA-dependent nucleic acid
polymerases are enzymes operating outside the "central dogma" and
hence are not ordinarily used by healthy cells.

For many of the RNA viruses and some DNA viruses, the cellular
membrane undergoes modification by incorporating viral enzymes.
This membrane becomes wrapped around an assembled particle to form a
lipoprotein envelope which is pinched off and released into its sur-
roundings as a virion.

Two additional features of virus-host interactions are *latency*
and *transformation*. The genomes (or DNA copies of them) of herpes
viruses and oncoviruses are incorporated into the host's chromosomes
and replicated with the latter in perpetuity. Stress--chemical,
physical, or emotional--can reactivate the latent genes, and this
causes a new outbreak of the disease; for example, cold sores (herpes simplex virus). Oncovirus infection of a cell may not lead to
production of new viruses but to a transformation of the cell into
a malignant cell, causing cancer.

For a fuller introduction to virology see, for example, Ref. 3.

3. INTERFERING WITH THE VIRAL CYCLE

The intimate relationship between host cell and virus makes it difficult to find a substance which inhibits the virus without being
toxic to the cell.

Far fewer substances are effective in vivo (in animals or humans) than in vitro (cell culture) because of the added requirements that they must be soluble in body fluids, find the infected cell, and be capable of penetrating it. Furthermore, viral multiplication may be limited to the initial site (as in most respiratory infections) or may spread to secondary sites via the bloodstream or other routes.

Unfortunately, the symptoms of a viral infection are caused not only by viral cell damage but also by host responses. Overreaction of the immune system can be responsible for a disease (e.g., viral encephalitis) in which damage by the virus itself may be minimal. In the many cases where the symptoms appear after the peak of virus multiplication, antiviral chemotherapy is meaningless and general supportive measures need to be taken [4].

TABLE 2

Some Antiviral Agents

Step in the viral cycle	Antiviral agent
Attachment	Immunoglobulins
Penetration	α-Aminoadamantane (amantadine)
Uncoating	Immunoglobulins
Replication	Nucleoside analogs, e.g., Ara-C
(DNA)	α-(N)-Heterocyclic thiosemicarbazones Phosphonoacetic acid Zinc ions
(RNA)	Rifamycin 2-(α-Hydroxybenzyl)-benzimidazole Guanidine/α-amanitin
Transcription	Actinomycin D/interferon Isatin β-thiosemicarbazone
Translation	Puromycin/interferon
Assembly	Isatin β-thiosemicarbazone Zinc ions/rifamycin

Source: Compiled from data given in Refs. 3, 4, and other references given in the text.

Despite these limitations, antiviral drugs are known, as shown
in Table 2. The activity of many current antiviral agents can be
related to steps in the viral cycle. Virion-associated and virus-
induced enzymes are often very different in their properties from
enzymes of host cells [5]. One aspect which has been little ex-
plored is the contribution that might be made by metal ions and
chelating agents in this context.

4. CHELATION AND BIOLOGICAL ACTIVITY

Alkaline and alkaline earth cations and trace amounts of first-row
transition metal ions (from manganese to zinc) are essential to the
well-being of living animals [6]. The major part of the transition
metal ions is bound to proteins. The metal ion may be an integral
part of a metalloenzyme (which is a metal-apoprotein complex), in
which case it is usually bound at the active site of the enzyme. It
may also stabilize the structures of proteins and nucleic acids.

Copper and iron enzymes are usually involved in redox reactions. Enzymes that contain zinc catalyze hydrolytic reactions or
the formation of amide and ester bonds. Thus it is likely that all
known nucleic acid polymerases are zinc enzymes [7].

Organisms can be regarded as finely tuned multimetal, multi-
ligand systems. If a heavy metal ion is present in excess, toxic
effects are produced, either because the wrong metal is bound to the
protein or because the right metal is present but it also binds to
other (inappropriate) sites in biomolecules. If the essential metal
ion is not present in sufficient amount, metalloenzymes cannot be
formed or else they undergo dissociation. An example of this biphasic character is provided by the lethal Menke's kinky hair syndrome and Wilson's disease (hepatolenticular degeneration). The
former results from the failure to absorb copper from the gut, while
the latter arises from excessive levels of exchangeable copper in
the blood.

Each tissue of an animal or plant contains many different kinds
and amounts of ligands, such as amino acids, peptides, and carboxylic

acids, together with small amounts of essential metal ions. The
mathematical treatment of such multimetal-multiligand systems has
been described [8], with particular reference to blood-plasma
equilibria. This approach, with better constants for the albumin
complexes [9], has been used to assess the importance of metal complexation of the antiviral drug phosphonoacetate under biological
conditions [10]. The equilibrium concentrations of the free metal
ions are expressed as pM values (= -log [M]). Thus in blood plasma,
with $[Cu]_{tot}$ = 1.1 μM and $[Zn]_{tot}$ = 15 μM, the computed values of pCu
and pZn are 15.5 and 8.8, respectively [9]. These values indicate
that polymeric species such as $Cu_2his_2(OH)_2$ which may be very significant under laboratory conditions are generally negligible in a
biological environment.

The total ligand concentrations in biological systems are much
larger than the total metal ion concentrations, so that mixed ligand-
transition metal complexes are extensively formed [11]. Especially
with copper(II) these mixed ligand complexes can be considerably
more stable than statistically expected, so that knowledge of the
corresponding stability constant is essential for a meaningful
simulation of the metal ion distribution in the blood plasma model.
The determination of these constants depends critically on the knowledge of the equilibria involving the parent binary complexes, precise measurements, and sophisticated computer programs. In much recent literature these requirements are not met. For example, although the main copper complex of dopa under titration conditions is
a dimeric complex [12], this species has been omitted from calculations of the stability constants for mixed ligand-dopa complexes
[13].

In comparing the relative stabilities of complexes under
specified conditions of pH, it may be quite misleading to use conventional stability constants. These express the complexing ability
of the free, unprotonated ligand, the equilibrium concentration of

which is critically dependent on the pH and the pK of the ligand.
For comparison at a given pH conditional ("apparent") stability
constants should be used [14]. In a conditional stability constant,
the free ligand concentration is replaced by the total concentration
of uncomplexed ligand, including its protonated forms.

The histidinato complexes of copper and zinc and the cysteinato
complexes of zinc are the major low-molecular-weight complexes of
these two metal ions in blood plasma. If a ligand forms less stable
copper and zinc complexes than does histidinate (as expressed in
their apparent stability constants), it is unlikely to be competitive
under biological conditions. As an example, the conditional stability constants for the zinc and copper complexes of dopa [12,15] are
smaller by 1 to 2 orders of magnitude than those of the histidinato
complexes, casting doubt on the often-emphasized importance of metal
ion chelation by dopa. It is worth remembering that most metal-
ligand equilibria are attained rapidly. Hence it is not important
whether a "preformed complex" or an appropriate mixture of the components is used.

Although, among first-row transition metal ions, copper complexes are almost always the most stable, it does not follow that
the addition of a strong ligand to a multimetal-multiligand system
will favor complex formation with copper. Differences in relative
stabilities are also important. Thus the EDTA complexes of copper
are considerably more stable than those of zinc or calcium, but
addition of EDTA to blood plasma considerably alters calcium and
zinc distribution but does not result in very much copper-EDTA complex being formed [9].

Quantitative considerations such as these can provide a useful
guide to the possibilities of chelation therapy and to an understanding of the roles of metal ions and chelating agents in biological systems.

5. CHELATING AGENTS WITH ANTIVIRAL ACTIVITY

5.1. Thiosemicarbazones

The weak antiviral activity of benzaldehyde thiosemicarbazone and
its derivatives prompted structure-activity studies which culminated
in methisazone (1-methylisatin β-thiosemicarbazone, I) which, given

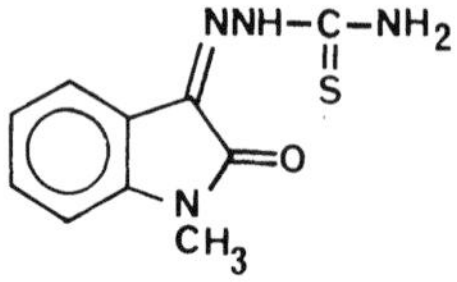

(I)

orally, was successfully used in the prophylaxis of smallpox
(variola) [2].   Isatin β-thiosemicarbazone and its N-methyl and N-
ethyl derivatives are active against a wide range of pox viruses.
Of these, methisazone has found medical application in treating
Eczema vaccinatum [16], *Vaccinia gangrenosum* [17], and other vac-
cinial complications [18].   Antiviral applications of thiosemicarbazones prior to 1970 have been reviewed [19].   Thiosemicarbazones
have been shown to be active against DNA and RNA viruses, tumors,
protozoa, and fungi, but there is often little margin between effective dose levels and toxicity.

As well as the thiosemicarbazones related to methisazone there
are two other types of thiosemicarbazones with biological activity,
namely the α-(N)-heterocyclic thiosemicarbazones (for example, 2-
formylpyridine thiosemicarbazone, II) and the bisthiosemicarbazones

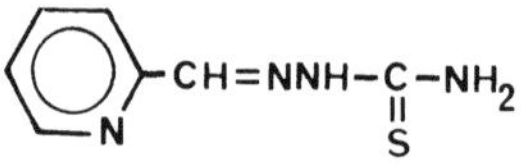

(II)

of α,β-dicarbonyl compounds (for example, kethoxal bisthiosemicarbazone, III).

(III)

    Deprotonation of thiosemicarbazones gives the tautomeric thio-
late form $-\overset{|}{\underset{}{C}}=\overset{*}{N}-N=C(NH_2)-\overset{*}{S}^-$. Crystal structure determinations showed
that they coordinate metal ions through the 1'-nitrogen and the sul-
fur (marked with an asterisk) [20]. Thiosemicarbazones containing
no additional donor atoms reduce copper(II) to copper(I), while bisthiosemicarbazones and tridentate thiosemicarbazones form stable
copper(II) complexes [20].

5.1.1. *Isatin β-Thiosemicarbazones*

    The inhibitory effects of isatin β-thiosemicarbazone (IBT) and
its derivatives on vaccinia virus are produced late in the viral
cycle. The assembly and maturation of the viral components is inter-
fered with because of a failure to generate "late" proteins [21].
The two main structural polypeptides are not available, and the assembly of viral cores is inhibited [22]. This is because polypeptide
4a fails to cleave from its (properly synthesized) precursor [23],
leaving the viral DNA uncoated and sensitive to the action of nucleases [22]. The activity of isatin β-thiosemicarbazone against vaccinia is lost if the sulfur atom is replaced by an oxygen atom or an
NH group [24,25].
    It has been suggested [26] that isatin β-thiosemicarbazones
act by metal ion binding, but this has been disputed [19]. Little
is known about the metal complexes of isatin β-thiosemicarbazones.
Metal complexes have been prepared [27,28] and some infrared spectral

data were given [28]. It was concluded that IBT in polar solvents
was in the E (anti) conformation [28]. Tomchin et al. [29] contra-
dicted this conclusion, and the spectroscopic assignments were
criticized by Campbell [20]. A pH-metric, kinetic, and nuclear
magnetic resonance (NMR) spectroscopic study of 5-sulfoisatin β-
thiosemicarbazone (SIBT) showed that in aqueous (and also DMSO)
solution the Z (syn) isomer predominates [30] (IVa). The deprotonated ligand is more stable in the E configuration, in which the 1'-
nitrogen and the 2-oxygen atoms are accessible for chelation (IVb).

(IV)

The conditional stability constants for the zinc complexes of
SIBT and its 1-methyl derivative are very similar: $\log \beta_1' = 4.5$ and
$\log \beta_2' = 9.0$ at pH 7.4 [31]. They are comparable with those of the
histidinato zinc complexes $\log \beta_1' = 4.8$ and $\log \beta_2' = 8.6$. The
iron(II) complexes of SIBT are weaker than those of zinc, while with
iron(III) ferric hydroxide is precipitated. Copper(II) is reduced
by these ligands, a portion of which is oxidized to the sulfonic
acid. The main complex formed at physiological pH is $Cu(I)_n L_n^{n-}$ (n
> 6) at total copper concentrations down to 8 μm. Monomeric 1:2
complexes were found only in presence of a very large excess of
ligand [32]. Similar reactions are likely with IBT and its 1-methyl
derivative, MIBT. The complex $Cu_n L_n$ would be uncharged and probably
insoluble in aqueous solution.

The formation of insoluble "ternary complexes" have been
claimed from filter retention and centrifugation experiments when

copper(II) and MIBT or a preformed copper(II)-MIBT 1:1 complex is
added to nucleic acids [33] or proteins [34]. This Cu(II)-MIBT
complex is rather insoluble in water, dissolves readily in dimethyl
sulfoxide, and has no counterion [35]. These rather puzzling properties suggest the formation of the Cu(I)MIBT complex, the "ternary
complexes" being a coprecipitate of two poorly soluble substances.
The interaction between CuMIBT and nucleic acids is reversed by adding DMSO or a chelating agent such as histidinate. Hence this co-
precipitation is unlikely to be significant under in vivo conditions,
but could well account for the observed [36] inhibition of many
viruses on extracellular contact with MIBT plus copper ions in the
absence of other chelating agents. This may also be true for the
synergism of MIBT and copper ions in the in vitro inhibition of Rous
sarcoma virus reverse transcriptase [37].

    5-Sulfoisatin β-semicarbazone (the oxygen analog of SIBT) is a
very weak ligand for copper and zinc, and 4-sulfobenzaldehyde thiosemicarbazone does not form stable zinc complexes below pH 7.5, at
which point zinc hydroxide precipitates. The differences in complexing ability of these ligands parallels the activity of the un-
sulfonated compounds against vaccinia virus. However, γ-thiochro-
manone 4-thiosemicarbazone (V) has the same mode of action against

(V)

vaccinia virus as IBT [38], although this compound would not be expected to form complexes stronger than benzaldehyde thiosemicarbazone.

    The reactions of sulfobenzaldehyde thiosemicarbazone and SIBT
with copper(II) are similar and indicate that the 2-oxygen contributes little to the coordination of copper ions. However, the mode
of action of IBT against vaccinia virus, by inhibition of the

cleavage of a precursor polypeptide, does not suggest the involvement of copper ions.

Isatin β-thiosemicarbazones inhibit, less efficiently, a range
of other DNA and also RNA viruses [39] but show different structure-
activity relationships. In isatin β-isothiosemicarbazones with an
alkyl substituent at the side chain sulfur atom [-Ċ=N-N=C(NH$_2$)-S-R]
the sulfur atom has no significant complexing ability. Nevertheless,
they inhibit Mengo virus RNA synthesis [40] and the corresponding RNA
polymerase [41]. Some copper and zinc complexes have been prepared,
and complexation has been claimed to be important for their biological activity [42]. However, precipitation of complexes from concentrated solutions must not be taken as a measure of their stabilities.

5.1.2. α-(N)-Heterocyclic Thiosemicarbazones

Thiosemicarbazones derived from 1-formylisoquinoline and 2-formyl-
pyridine and related heterocyclic bases (VI) inhibit the growth of

(VI)

DNA viruses of the herpes family [43] and Rous sarcoma virus [44].
They are also active against transplanted rodent neoplasma [45] and
canine lymphoma [46]: their antineoplastic activity has been reviewed [47]. This is related to their inhibition of DNA synthesis
by action on ribonucleoside diphosphate reductase. The water-
soluble 4-methyl-5-amino-1-formylisoquinoline thiosemicarbazone is
one of the most successful antineoplastic agents in this group [48],
but its antiviral activity has not been assessed.

The iron(II) enzyme ribonucleoside diphosphate reductase is
essential for the synthesis of deoxyribonucleoside diphosphates
which, in turn, are needed for the DNA synthesis. In enzyme assays,

it has been shown that α-(N)-heterocyclic thiosemicarbazones have some inhibitory activity, possibly by binding to the iron of the enzyme [49]. However, adding ferric ions or the iron complex of these ligands greatly enhanced the inhibitory activity, so that the iron complex was the actual inhibitor [49,50]. On the other hand, the copper complex of formylpyridine thiosemicarbazone was found not to dissociate in Ehrlich cells, and to be cytotoxic, possibly by acting as a catalyst for the oxidation of thiols by oxygen [51].

These ligands chelate through the ring nitrogen, the 1'-nitrogen, and the sulfur atoms [20]. Complexes with iron, cobalt, nickel, and copper have been synthesized [52,53]. The inhibitory activity on DNA synthesis parallels the expected chelating ability of these ligands. Thus, the seleno analog of 5-hydroxy-2-formylpyridine thiosemicarbazone was less active than the parent compound, and the guanylhydrazone and semicarbazone were inactive [54]. Conversion to the N oxide eliminated the activity of 6-formyl-purine thiosemicarbazone [55].

Stability constants for the 2-acetyl and 2-formylpyridine thiosemicarbazone complexes of copper and zinc are given in Table 3.

TABLE 3

Stability Constants of Metal Complexes of 2-Formyl
and 2-Acetylpyridine Thiosemicarbazone

	Cu		Zn	
	Formyl[a]	Acetyl[b]	Formyl[a]	Acetyl[b]
$\log K_1$	16.9	13.2[c]	9.2	10.6[c]
pK_a of MLH	2.4	--	--	4.7
pK_a of ML[d]	8.3	4.9	7.4	5.9
$\log K_1'$ (pH 7.4)	13.3	13.1	5.6	8.5

[a]Ref. 56; no 1:2 complexes included.

[b]Ref. 9; including data for 1:2 complexes; no experimental details given.

[c]Assuming pK of HL = 11.0[a].

[d]Forming MLOH.

Differences between the two sets of results far exceed effects due to the additional methyl group. The stability constants of the copper complexes were determined from the competitive equilibria

$$CuL^+ + en \rightleftharpoons CuLen^+$$
$$CuLen^+ + en \rightleftharpoons Cu(en)_2^{2+} + L^-$$

where HL = formylpyridine thiosemicarbazone and en = ethylenediamine [56]. A good value was obtained for log β_{CuLen} = 22.43 ± 0.02, but because CuL is not a major species under the experimental conditions the single-step constants log K_{CuL}^{Cu} and log K_{CuLen}^{CuL} show large deviations from the mean (16.6 to 17.1 and 5.3 to 5.8, respectively). In spite of experimental uncertainty, these ligands appear to complex copper ion more strongly than does albumin, and zinc more strongly than does histidinate under physiological conditions. Conditional constants of log β_2' = 21 [for FE(III)] and = 15.8 [for Fe(II)] at pH 7.4 have been reported for the 1:2 iron complexes of 2-formylpyridine thiosemicarbazone [57].

Other experiments showed clearly that these complexes are competitive in a biological environment. Formylpyridine thiosemicarbazone removes iron(III) from ferritin, while Fe(III)L$_2$ and Fe(II)L$_2$ are stable in plasma, the latter probably predominating in vivo [57]. Excretion of the iron(II) complex followed the administration of 5-hydroxy-2-formylpyridine thiosemicarbazone to patients with cancer [58,59]. There is evidence for a mixed ligand-copper(II) complex of formylpyridine thiosemicarbazone in plasma [57].

The antiviral activity of 2-acetylpyridine thiosemicarbazone, observed in an assay of influenza virus RNA-dependent RNA polymerase activity might be due to chelation of the zinc ion in the enzyme [60]. The antiviral activity of α-(N)-heterocyclic isothiosemicarbazones (substituted at the side chain sulfur) is unlikely to be due to chelation, especially in view of the greater activity of a 4-formyl pyridine isothiosemicarbazone compared to the 2-derivative [61]. Phosphorous analogs of 2-formylpyridine thiosemicarbazone [in which -C(NH$_2$)=S is replaced by -P(R$_2$)=S] had antitumor activity when combined with copper but not with ferrous ion [62].

5.1.3. Bis-Thiosemicarbazones

Kethoxal bisthiosemicarbazone (H_2KTS, 3-ethoxy-2-oxobutyraldehyde bisthiosemicarbazone, III) was active against vesicular stomatitis virus in chick embryo cells by inhibiting viral mRNA and protein synthesis [63], but there was also marked inhibition of cellular DNA, RNA, and protein synthesis. Methisazone and 2-formylpyridine thiosemicarbazone did not inhibit this virus.

Pyruvaldehyde and kethoxal bisthiosemicarbazones are carcinostatic, and dietary copper enhances the anticancer activity of the parenterally administered drug [64]. The copper chelate was the active species for KTS [65], possibly as an oxidizing agent which interferes with the energy transport system [66]. Copper(II) complexes of 1,2-bisthiosemicarbazones readily undergo oxidation and reduction [67].

These 1,2-bisthiosemicarbazones form very stable complexes with metal ions. The conditional stability constant of the copper complex of KTS at pH 7.4 has been reported to be $10^{18 \cdot 4}$ [68], and this complex does not dissociate appreciably in human plasma and mouse ascites fluid [68]. The corresponding constant for the zinc complex is claimed to be smaller by 12 orders of magnitude, but this is doubtful because of uncertainties in the method used to obtain it.

1,5-Bisthiosemicarbazones have antitumor activity, but their metal-binding ability is substantially less than for 1,2-bisthiosemicarbazones. The copper chelates of 1,5-bisthiosemicarbazones readily penetrated leukemia L1210 cells where they dissociated [69]. The drug then diffused from the cells and was available to shuttle more metal ions inward, causing cytotoxicity due to copper poisoning. This mechanism is unlikely to operate with 1,2- or 1,3-bisthiosemicarbazones because the metal complexes are too stable.

5.2. Phosphonoacetic and Phosphonoformic Acids

Phosphonoacetic acid (PAA, VII) has antiviral activity against all herpes viruses so far examined [70]. Administered orally or

$$\overset{\overset{\displaystyle O}{\parallel}}{\underset{\underset{\displaystyle O^-}{|}}{^-O-P-CH_2-C}}\overset{\diagup O}{\diagdown O^-}$$

(VII)

topically to mice experimentally infected with herpes simplex virus, PAA (as the disodium salt) significantly reduced the mortality associated with the infection [71]. The most valuable results with PAA have been in the treatment of herpetic keratitis and iritis in eyes of rabbits [72,73]. In comparison with other antiherpes compounds, PAA had the best therapeutic activity [74]. However, treatment of established infections in vivo is not always positive [75]. PAA is of low systemic toxicity, but is often locally irritating. The existence of PAA-resistant mutants further limits the usefulness of this drug [76].

Phosphonoformate (PFA) is the only analog of PAA that shows comparable antiherpes activity [77], whereas a range of other analogs containing two groups such as SO_3^-, COO^-, COOR, PO_3^{2-}, PO_2OR^-, and $PO(OR)_2$ were inactive [78]. PFA has the advantage of not being locally irritant [79,80].

The mode of action of PAA on herpes simplex virus is the inhibition of herpes simplex virus DNA synthesis [81]. Replication of vaccinia virus is also inhibited by PAA, but much less strongly. Other DNA-containing viruses (simian virus-40 and human adenovirus-12) are not inhibited, nor are the RNA-containing viruses such as polio virus, rhinovirus, or measles virus [9]. PFA inhibits influenza virus RNA polymerase much more strongly than does PAA [82]; this activity was enhanced in the presence of magnesium or manganese-(II) ions.

Using the plaque reduction technique on cell cultures, PAA at 25-100 μg/ml has no obvious cytotoxic effect on uninfected cells in culture but effectively blocks herpes virus replication. PAA does not inhibit the synthesis of RNA and protein in infected cells, but at higher concentrations (200-1,000 μg/ml) in normal or infected

cells it inhibits cell growth and cellular DNA synthesis and hence
is cytotoxic [83,84]. The sensitive components are the DNA polymerases. Herpes viruses induce a virus-specific DNA polymerase
which is sensitive to PAA, with inhibition constants around 1-2 μM.
Eukariotic DNA polymerase-α is 15-30 times less sensitive and the
DNA polymerases-β and -γ are insensitive [85].

PAA blocks the site that should accept pyrophosphate liberated
in the formation of DNA from deoxyribonucleoside triphosphates [78,
86]. Thus, PAA and pyrophosphate inhibit herpes virus-induced DNA
polymerase in vitro in an analogous manner (but pyrophosphate is only
about 1/1000 as active). Both are noncompetitive inhibitors of deoxyribonucleoside triphosphates, whereas PAA is a competitive inhibitor of pyrophosphate in the pyrophosphate-deoxyribonucleoside
triphosphate exchange reaction. PAA-resistant mutants of herpes
viruses have been studied, and the DNA polymerase induced by the
mutant viruses is also PAA-resistant [87]. PAA-resistant DNA polymerase is also resistant to PFA [77] so that the two compounds apparently have the same mode of action [88].

The stability constants of the complexes of PAA with Mg^{2+},
Ca^{2+}, Cu^{2+}, and Zn^{2+} have been determined [10]. A computer-based
simulation of metal-ligand equilibria in blood plasma showed that
PAA did not significantly affect the copper ion distribution. Zinc/
PAA complexes were predominant among the low-molecular-weight zinc
complexes. The free magnesium ion concentration was significantly
reduced and appreciable amounts of CaPAA⁻ were formed. From the
latter result, it is not surprising that PAA is retained in the
bones [89].

From the chemical literature the 5-membered metal-chelate ring
formed by PFA would have a stability constant similar to, or slightly
greater than, the corresponding PAA complex. The 7-membered ring
that would be required for phosphonopropanoate would be unlikely to
contribute much stability to the metal complex. Phosphonopropanoate
is not active against herpes virus DNA polymerase [78]. As the antiviral results parallel the expected complexing abilities of these

ligands, and zinc seems to be bound by PAA under simulated in vivo
conditions, we could speculate that PAA is bound to the DNA polymerase through the zinc ion of this enzyme.

5.3. β-Diketones

Acetylacetone (pentan-2,4-dione) is a well-known chelating agent.
Many 4-substituted 3,5-heptanediones, derivatives of acetylacetone,
have antiviral activity [see Refs. 90 and 91 and references therein].
They are active against herpes simplex viruses and also some RNA
viruses (rhinoviruses, influenza virus) in vitro. Arildone, 4-[6-
(2-chloro-4-methoxy)phenoxyl]hexyl-3,5-heptanedione (VIII), is a

(VIII)

promising member of this series. It inhibits herpes simplex virus
in tissue culture if it is added to the culture within 6 hr of infection [92]. From results with polio virus, it is suggested that
uncoating of the virion is inhibited, so that the virus-induced
shutoff of host cell protein synthesis is prevented [93]. Arildone
is not virucidal and the inhibition can be removed by washing. It
does not interfere significantly with the macromolecular synthesis
in the host cells [92]. Arildone [94] and other 3,5-heptanediones
[90,91] have been effective topically against herpes virus infections in the rabbit eye, guinea pig skin, and mouse vagina.

     Our computer-based blood plasma model shows that there would
be a significant formation of copper acetylacetonate complexes
among the low-molecular-weight complexes at ligand concentrations

above 10^{-4} M. Complexation of iron by a β-diketonate also seems
plausible, but currently the extent of complexation and the mode of
action of the β-diketonates are unknown.

5.4. Other Chelating Agents

A discussion of chelating agents that inhibit isolated viral enzymes
in vitro is beyond the scope of this review. In particular, it
seems that any sufficiently strong chelating agent inhibits viral
and also cellular DNA or RNA polymerases. Hence we limit our selection to ligands that have shown antiviral activity either in vivo or
in tissue culture. These substances almost certainly form complexes
in a biological environment.

5.4.1. *8-Hydroxyquinoline*

In the presence of metal ions this ligand shows antifungal and antibacterial activity. It inhibited RNA-dependent DNA polymerase of
Rous sarcoma virus and abolished the ability of this virus to transform cells [95]. This polymerase was also inhibited by 8-mercapto-
and 8-amino-quinoline and by 8-hydroxyquinoline-5-sulfonate. The
last-named and its metal complexes are readily water-soluble. The
conditional stability constants at pH 7.4 of the copper and zinc
complexes are larger than those of the histidinato complexes by 2-3
logarithmic units. Hence complex formation in vivo can be considered
certain. Substitution of the hydroxyl group by a mercapto group enhances the stability of the zinc complexes much more than that of the
copper complexes, while 8-aminoquinoline is a much weaker ligand for
both metal ions.

5.4.2. *Flavonoids*

Flavonoids were virucidal to enveloped viruses [96]. Quercetin and
morin had a prophylactic effect against rabies virus in mice [97]
and were also effective against Mengo virus-induced encephalitis in

mice [98]. Flavonoids also showed activity against vesicular stomatitis virus when tested in cell culture [99]. Maximum inhibition
was at 200 µg flavonoid/ml, added 6-8 hr before viral infection.
Complex formation may take place through the carbonyl group and the
deprotonated perihydroxyl group, as in complex formation with alizarin red S. At pH 7.4, the zinc complex of alizarin red S is an
order of magnitude more stable than the zinc histidinate complex,
whereas the copper complex is less stable than histidinato copper.
Assuming comparable stability for the complexes of the flavonoids,
complexation of zinc under biological conditions seems likely. Gly-
cosidic anthracycline antibiotics such as adriamycin and daunomycin
contain a flavonoid-type chelating group. They inhibit, in vitro,
the DNA polymerases of retroviruses [100,101] and decrease the number of sarcomas induced by Rous sarcoma virus in chicks [102]. The
currently suggested mode of action, by specific interaction with AT
pairs on the nucleic acid templates [101,103], makes involvement of
metal ions unlikely.

5.4.3. Bleomycins

Bleomycins are a family of glycopeptide antibiotics that are usually
isolated as their copper complexes. Copper-free bleomycins inhibit
vaccinia virus replication in HeLa cells and protect mice against
vaccinia infection [104]. The significance of metal complexation in
the biological activity of bleomycins has been reviewed [105].

5.4.4. EDTA

EDTA inactivated purified tobacco rattle virus, possibly by removing
calcium ions bound to the virion [106]. Calcium ions appear to be
bound by a number of virions and their removal facilitates the
separation of the proteins from the nucleic acid [107]. Thus, the
low intracellular concentration of calcium may lead to a dissociation
of the calcium from virions and thereby trigger the uncoating processes [108]. If the virus is disassembled extracellularly, the
nucleic acid is accessible to attack by nucleases [108]. However,

the ubiquitous nature of calcium makes it difficult to explore calcium chelation as a therapeutic avenue.

Calcium ion is also an essential cofactor for all neuramini-
dases. EDTA was effective against influenza virus neuraminidase
[109]. Hence, it seems possible that application of EDTA solutions
by means of a nasal spray might slow down the release and spread of
the virus, giving the host's immune system time to react.

5.4.5. Selenocystine

Selenocystine inhibited influenza virus PR8 in mice and Rous sarcoma
virus in chicks by its selective action on the viral RNA polymerase;
it did not affect the DNA-dependent RNA polymerase [110].

5.4.6. Other Complexing Agents

The glyoxal derivative (IX), 6-bromonaphthoquinone (Bonaphthon, X),
(XI) [111], the tetrazine humic acids and polyphenols [112] were

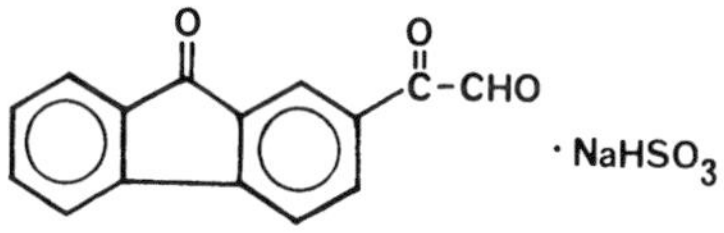

(IX)

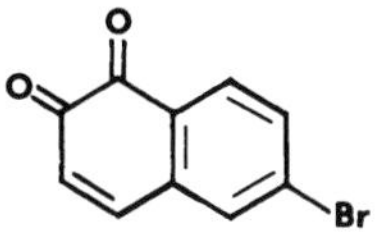

(X)

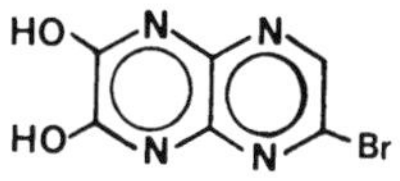

(XI)

active against herpes viruses. A common feature of many antiherpes
agents, including PAA, PFA, and β-diketones, is that they are biden-
tate ligands with two oxygens as metal-binding atoms.

Many other antiviral compounds, in particular the nucleoside
analogs, bind metal ions in pure solutions. However, under physiological conditions they are not likely to be able to compete successfully for the low equilibrium concentrations of metal ions.

6. DELIVERY OF DRUGS TO TARGETS

Many difficulties beset the application of drugs, active in vitro, to living organisms. As in the case of thiosemicarbazones, sparing solubility in water may militate against oral administration, necessitating the synthesis of structurally modified agents in which hydrophilic groups such as amino and hydroxyl have been inserted. If the agent carries a high charge, it may not penetrate cell barriers: this limits the uptake of ions such as $Ca(EDTA)^{2-}$ or $H(EDTA)^{3-}$. Neutrality may not be a requirement for penetration: Thus Petering [113] found that although a compound had a 1-octanol/H_2O partition coefficient of about 5, it was not readily absorbed by Ehrlich cells, whereas its copper complex (carrying a +1 charge) and two iron complexes were readily taken up.

The attainable level of a drug in any body tissue also depends on the rate of excretion via the kidney and the ease with which the agent undergoes degradation in the body. A combination of these factors may lead to the failure to reach pharmacologically active concentrations. Many of these difficulties might be overcome by using a lipid-soluble prodrug which was metabolized in the target cells to give the active form.

Alternatively it might be possible to administer a drug encapsulated in liposomes which, because of their phospholipid exterior, can pass through cell barriers and reach a selected target.

Liposomes have been used to inject DTPA into mice for the removal of a fraction of the metal plutonium that was not accessible to the usual DTPA therapy [114]. High intracellular levels of EDTA and DTPA were maintained when liposomes containing these chelating agents were injected intravenously [115]. The tissue distribution of liposomes containing EDTA varied with the charge, if any, on the liposomes [116] and with their size [117].

Promising results were obtained when liposomes containing cytotoxic agents were used to treat mice with ascites tumors [118,119]. An unsuccessful attempt to inhibit influenza viral activity in ferrets using liposome-encapsulated CaEDTA and CaDTPA has been described [120]. Erythrocyte "ghosts" (made by the rapid lysis of red cells) may also find application in carrying therapeutic agents to the liver or spleen. Modification of the "ghost" surfaces might lead to improved specificity [121,122].

For recent reviews of liposomes and their applications, see Refs. 123 and 124.

7. METAL IONS IN ANTIVIRAL CHEMOTHERAPY

Metal ions are essential to the proper functioning of living cells, the levels of free metal ions being controlled by the complexing agents that are also present. If the optimum pM values of a virus lie outside the ranges found in normal cells, the possibility exists that viruses might be inhibited by adding excess metal ion. Metal ions may bind to inappropriate sites of enzymes or to the substrate. (Excess of a metal ion favors binary complex formation instead of ternary complexes such as apoenzyme-zinc substrate.)

Excess zinc ion (0.2 mM) inhibits the herpes simplex virus DNA polymerase in vitro [125] but does not affect cell DNA synthesis [126]. The sensitivity of herpes simplex DNA synthesis to zinc ion makes it possible to treat herpes keratitis in humans with 0.5% zinc sulfate [127]. Conversely, deficiency of zinc ion leads to virus inhibition. However, as wound healing is facilitated if adequate zinc levels are maintained, it might be preferable to use a lipid-soluble zinc salt for the topical treatment of cutaneous herpetic lesions rather than phosphonoacetate.

Solution 0.5 mM in zinc ions interferes with the proteolytic cleavage of large polypeptides to capsid polypeptides in the replication of some RNA viruses [128] including foot-and-mouth disease virus [129], human rhinovirus, and polio virus [130,131]. It may be that binding of zinc ions to the polypeptide prevents access of the protease which would normally cleave it [132].

The toxicity of zinc ions for many kinds of bacteriophages and animal viruses might be used to block certain stages in their replication [128]. Metal ions inhibit the activity of the RNA-dependent DNA polymerase of Rous sarcoma virus and its cell-transforming ability [133].

Therapy with heavy metal ions other than zinc is severely restricted because of their toxicity. However, subcutaneous injection of copper(II), rhodium, and iridium salts into mice infected with neurovaccinia had a therapeutic effect [134].

ABBREVIATIONS

DMSO	Dimethylsulfoxide
DNA	Deoxyribonucleic acid
DTPA	Diethylenetriamine pentaacetate
EDTA	Ethylenediamine-N,N,N',N'-tetraacetate
HEDTA	Hydroxyethylethylenediamine triacetate
his	Histidinate
IBT	Isatin-β-thiosemicarbazone
KTS	Kethoxal bisthiosemicarbazone
	(3-ethoxy-2-oxobutyraldehyde bisthiosemicarbazone)
L	General for ligand
M	General for metal ion
MIBT	Methisazone, 1-methylisatin β-thiosemicarbazone
mRNA	Messenger RNA
PAA	Phosphonoacetic acid
PFA	Phosphonoformic acid
pK	$-\log([X]\cdot(H)/[HX])$
RNA	Ribonucleic acid
SIBT	5-Sulfonatoisatin β-thiosemicarbazone
β_n	Cumulative stability constant: $[ML_n]/([M]\cdot[L]^n)$
β'_n	Conditional ("apparent") stability constant

REFERENCES

1. A. Albert, *Selective Toxicity*, 6th ed., Chapman and Hall, London, 1979.

2. D. J. Bauer, L. St. Vincent, L. H. Kempe, and A. M. Downie, *Lancet, 2,* 494 (1963).

3. F. D. Fenner and D. O. White, *Medical Virology*, Academic Press, London, 1976.

4. H. Stalder, *Yale J. Biol. Med., 50,* 507 (1977).

5. S. Kit, *Pharmacol. Ther., 4,* 501 (1979).

6. E. J. Underwood, *Trace Elements in Human and Animal Nutrition,* 4th ed., Academic Press, New York, 1977.

7. J. F. Riordan and B. L. Vallee, *Trace El. Hum. Health Dis., 1,* 227 (1976).

8. D. D. Perrin and R. P. Agarwal, *Metal Ions Biol. Syst., 2,* 168 (1973).

9. R. P. Agarwal and D. D. Perrin, *Agents Actions, 6,* 667 (1976).

10. H. Stünzi and D. D. Perrin, *J. Inorg. Biochem., 10,* 309 (1979).

11. For a fuller discussion, see Vol. 2 of this series.

12. A. Gergely and T. Kiss, *Inorg. Chim. Acta, 16,* 51 (1976); J. E. Gorton and R. F. Jameson, *J. Chem. Soc., A,* 2615 (1968).

13. K. S. Rajan, S. Mainer, and J. M. Davies, *Bioinorg. Chem., 9,* 187 (1978); *J. Inorg. Nucl. Chem., 40,* 2089 (1978).

14. For a discussion of conditional stability constants, see A. Ringbom, *Complexation in Analytical Chemistry,* Interscience, New York, 1963.

15. A. Gergely, T. Kiss, and G. Deák, *Inorg. Chim. Acta, 36,* 113 (1979).

16. W. Turner, D. J. Bauer, and R. H. Nimmo-Smith, *Brit. Med. J., 1,* 1317 (1962).

17. D. J. Bauer, *Ann. N.Y. Acad. Sci., 130,* 110 (1965).

18. D. M. McLean, *Ann. N.Y. Acad. Sci., 284,* 118 (1977).

19. D. J. Bauer, in *Chemotherapy of Virus Diseases, International Encyclopedia of Pharmacology and Therapeutics* (D. J. Bauer, ed.), Sec. 61, Vol. 1, Pergamon Press, Oxford, 1973.

20. M. J. M. Campbell, *Coord. Chem. Rev., 15,* 279 (1975).

21. T. H. Pennington, *J. Gen. Virol., 35,* 567 (1977).

22. E. Katz, E. Margalith, and B. Winer, *J. Gen. Virol., 40,* 595 (1978).

23. E. Katz, E. Margalith, B. Winer, and N. Goldblum, *Antimicrob. Agents Chemother.*, *4*, 42 (1973).

24. R. L. Thompson, S. A. Minton, J. E. Officer, and G. H. Hitchings, *J. Immunol.*, *70*, 229 (1953).

25. F. W. Sheffield, D. J. Bauer, and S. M. Stephenson, *Brit. J. Exp. Path.*, *41*, 638 (1960).

26. D. J. Bauer, *Brit. J. Exp. Path.*, *36*, 105 (1955).

27. V. Hovorka and Z. Holzbrecher, *Collect. Czech. Chem. Commun.*, *14*, 248 (1949).

28. P. Barz and H. P. Fritz, *Z. Naturforsch.*, *25B*, 199 (1970).

29. A. B. Tomchin, I. S. Ioffe, A. I. Kol'tsov, and Yu. V. Lepp, *Chem. Heterocycl. Comp.* (Engl. trans.) 4, 437 (1974).

30. H. Stünzi, *Aust. J. Chem.*, *34*, 373 (1981).

31. H. Stünzi, Proc. 21st Int. Conf. Coord. Chem., Toulouse, 173 (1980).

32. H. Stünzi, *Aust. J. Chem.*, in press (1981).

33. P. E. Mikelens, B. A. Woodson, and W. Levinson, *Biochem. Pharmacol.*, *25*, 821 (1976).

34. W. Rohde, R. Shafer, J. Idriss, and W. Levinson, *J. Inorg. Biochem.*, *10*, 183 (1979).

35. W. C. Kaska, C. Carrano, J. Michalowski, J. Jackson, and W. Levinson, *Bioinorg. Chem.*, *8*, 225 (1978).

36. W. Levinson, V. Coleman, B. Woodson, A. Rabson, J. Lanier, J. Witcher, and C. Dawson, *Antimicrob. Agents Chemother.*, *5*, 398 (1974); J. C. Logan, M. P. Fox, J. H. Morgan, A. M. Makahon, and C. J. Pfau, *J. Gen. Virol.*, *28*, 271 (1975); M. P. Fox, L. H. Bopp, and C. J. Pfau, *Ann. N.Y. Acad. Sci.*, *284*, 533 (1977).

37. W. Levinson, A. Faras, P. Woodson, J. Jackson, and J. M. Bishop, *Proc. Nat. Acad. Sci. U.S.*, *70*, 164 (1973).

38. E. Katz, E. Margalith, and B. Winer, *J. Gen. Virol.*, *25*, 239 (1974).

39. D. J. Bauer, K. Apostolov, and J. W. T. Selway, *Ann. N.Y. Acad. Sci.*, *173*, 314 (1970).

40. M. Tonew and E. Tonew, *Antimicrob. Agents Chemother.*, *5*, 393 (1974).

41. E. Tonew, G. Löber, and M. Tonew, *Acta Virol.*, *18*, 185 (1974).

42. L. Heinisch and D. Tresselt, *Pharmazie*, *32*, 582 (1977).

43. W. Brockman, R. W. Sidwell, G. Arnett, and S. Shaddix, *Proc. Soc. Exp. Biol. Med.*, *133*, 609 (1970).

44. W. Levinson, W. Rohde, P. Mikelens, J. Jackson, A. Antony, and T. Ramakrishnan, *Ann. N.Y. Acad. Sci.*, *284*, 525 (1977).

45. F. A. French and E. J. Blanz, *J. Med. Chem.*, *9*, 585 (1966).

46. W. A. Creasy, K. C. Agrawal, R. L. Capizzi, K. K. Stinson, and A. C. Sartorelli, *Cancer Res.*, *32*, 565 (1972).

47. K. C. Agrawal and A. C. Sartorelli, *Prog. Med. Chem.*, *15*, 321 (1978).

48. K. C. Agrawal, P. D. Mooney, and A. C. Sartorelli, *J. Med. Chem.*, *19*, 970 (1976).

49. L. A. Saryan, E. Ankel, C. Krishnamurti, and D. H. Petering, *J. Med. Chem.*, *22*, 1218 (1979).

50. K. C. Agrawal, B. A. Booth, E. C. Moore, and A. C. Sartorelli, *Proc. Amer. Assoc. Cancer Res.*, *15*, 73 (1974).

51. J. M. Knight, E. Whelan, and D. H. Petering, *J. Inorg. Biochem.*, *11*, 327 (1979).

52. A. V. Ablov and N. I. Belichuk, *Zh. Neorg. Khim.*, *14*, 179 (1969).

53. W. E. Antholine, J. M. Knight, and D. H. Petering, *J. Med. Chem.*, *19*, 339 (1976).

54. R. L. Michaud and A. C. Sartorelli, Abstr. Amer. Chem. Soc. 155th Nat. Meet., San Francisco, April 1968, N-54.

55. A. Giner-Sorolla, M. McCravey, J. Longley-Cook, and J. H. Burchenal, *J. Med. Chem.*, *16*, 984 (1973).

56. W. E. Antholine, J. M. Knight, and D. H. Petering, *Inorg. Chem.*, *16*, 569 (1977).

57. W. E. Antholine, J. Knight, E. Whelan, and D. H. Petering, *Mol. Pharmacol.*, *13*, 89 (1977); E. Ankel and D. H. Petering, *Biochem. Pharmacol.*, *29*, 1833 (1980).

58. R. C. DeConti, B. R. Toftness, K. C. Agrawal, R. Tomchick, J. A. R. Mead, J. R. Bertino, A. C. Sartorelli, and W. A. Creasey, *Cancer Res.*, *32*, 1455 (1972).

59. I. M. Krakoff, E. Etcubanas, C. Tan, K. Mayer, V. Bethune, and J. H. Burchenal, *Cancer Chemother. Rep.*, *58*, 207 (1974).

60. J. S. Oxford, in *Chemoprophylaxis and Virus Infection of the Respiratory Tract,* Vol. 1 (J. S. Oxford, ed.), CRC Press, Cleveland, 1977, Chap. 5.

61. L. Heinisch, M. Tonew, and E. Tonew, *Pharmazie, 32*, 752 (1977).

62. L. A. Cates, Y. M. Cho, L. K. Smith, L. Williams, and T. L. Lemke, *J. Med. Chem.*, *19*, 1133 (1976).

63. W. Levinson, H. Opperman, and J. Jackson, *J. Gen. Virol.*, *37*, 183 (1977).

64. J. G. Cappuccino, S. Banks, G. Brown, M. George, and G. S. Tarnowski, *Cancer Res.*, *27*, 968 (1967).

65. J. A. Crim and H. G. Petering, *Cancer Res.*, *27*, 1278 (1967).

66. D. H. Petering, *Bioinorg. Chem.*, *1*, 273 (1972).

67. L. E. Warren, S. M. Horner, and W. E. Hatfield, *J. Amer. Chem. Soc.*, *94*, 6392 (1972).

68. D. H. Petering, *Biochem. Pharmacol.*, *23*, 567 (1974).

69. D. Kessel and R. S. McElhinney, *Mol. Pharmacol.*, *11*, 298 (1975).

70. J. Hay, S. M. Brown, A. T. Jamieson, F. J. Rixon, H. Moss, D. A. Dargan, and J. H. Subak-Sharpe, *J. Antimicrob. Chemother.*, *Suppl. A.*, *3*, 63 (1977).

71. N. L. Shipkowitz, R. R. Bower, R. N. Appell, C. W. Nordeen, L. R. Overby, W. R. Roderick, J. B. Schleicher, and A. M. Von Esch, *Appl. Microbiol.*, *26*, 264 (1973).

72. R. F. Meyer, E. D. Vanell, and H. E. Kaufman, *Antimicrob. Agents Chemother.*, *9*, 308 (1976).

73. Y. J. Gordon, M. Lahar, S. Photiou, and Y. Becker, *Brit. J. Ophthal.*, *61*, 506 (1977).

74. S. Alenius and B. Öberg, *Arch. Virol.*, *58*, 277 (1978); J. Descamps, E. DeClerc, P. J. Barr, A. S. Jones, R. T. Walker, P. F. Torrence, and D. Shugar, *Antimicrob. Agents Chemother.*, *16*, 680 (1979).

75. S. Alenius and H. Nordlinder, *Arch. Virol.*, *60*, 197 (1979).

76. A. A. Newton, *Advan. Ophthal.*, *38*, 267 (1979).

77. J. M. Reno, L. F. Lee, and J. A. Boezi, *Antimicrob. Agents Chemother.*, *13*, 188 (1978).

78. S. S. Leinbach, J. M. Reno, L. F. Lee, A. F. Isbell, and J. A. Boezi, *Biochemistry*, *15*, 426 (1976).

79. S. Alenius, Z. Dinter, and B. Öberg, *Antimicrob. Agents Chemother.*, *14*, 408 (1978).

80. S. Alenius, *Arch. Virol.*, *65*, 149 (1980).

81. L. R. Overby, E. E. Robishaw, J. B. Schleicher, A. Rueter, N. L. Shipkowitz, and J. C. H. Mao, *Antimicrob. Agents Chemother.*, *6*, 360 (1974).

82. S. Stridh, E. Helgstrand, B. Lannerö, A. Misiorny, G. Stening, and B. Öberg, *Arch. Virol.*, *61*, 245 (1979).

83. E. S. Huang, *J. Virol.*, *16*, 1560 (1973).

84. O. Nyormoi, D. A. Thorley-Lawson, J. Elkington, and J. L. Strominger, *Proc. Nat. Acad. Sci. U.S.*, *73*, 1745 (1976).

85. C. L. K. Sabourin, J. M. Reno, and J. A. Boezi, *Arch. Biochem. Biophys.*, *187*, 96 (1978).

86. J. C. H. Mao and E. E. Robishaw, *Biochemistry*, *14*, 5475 (1975).

87. D. J. M. Purifoy and K. L. Powell, *J. Virol.*, *24*, 470 (1977).

88. E. R. Kern, L. A. Glasgow, J. C. Overall, Jr., J. M. Reno, and
     J. A. Boezi, *Antimicrob. Agents Chemother., 14,* 817 (1978).

89. B. A. Bopp, C. B. Estep, and D. J. Anderson, *Fed. Proc., 36,*
     939 (1977).

90. G. D. Diana, P. M. Carabateas, U. J. Salvador, G. L. Williams,
     E. S. Zalay, F. Pancic, B. A. Steinberg, and J. C. Collins,
     J. Med. Chem., 21, 689 (1978).

91. G. D. Diana, P. M. Carabateas, R. E. Johnson, G. L. Williams,
     F. Pancic, and J. C. Collins, *J. Med. Chem., 21,* 889 (1978).

92. M. F. Kuhrt, M. J. Fancher, V. Jasty, F. Pancic, and P. E.
     Came, *Antimicrob. Agents Chemother., 15,* 813 (1979).

93. J. J. McSharry, L. A. Caliguiri, and H. J. Eggers, *Virology,
     97,* 307 (1979).

94. F. Pancic, B. Steinberg, G. Diana, W. Gorman, and P. Came,
     17th Intersci. Conf. Antimicrob. Agents Chemother., New York,
     Oct. 12-14, 1977, Abstr. No. 289 (quoted in *Annual Reports in
     Med. Chem.,* 1978, p. 145).

95. W. Rohde, P. Mikelens, J. Jackson, J. Blackman, J. Whitcher,
     and W. Levinson, *Antimicrob. Agents Chemother., 10,* 234
     (1976).

96. I. Beladi, R. Pusztai, I. Mucsi, M. Bakay, and M. Gabor, *Ann.
     N.Y. Acad. Sci., 284,* 358 (1977).

97. W. C. Cutting, R. H. Dreisbach, and B. J. Neff, *Stanford Med.
     Bull., 7*(2), 137 (1949); W. C. Cutting, R. H. Dreisbach, and
     F. Matsushima, *Stanford Med. Bull., 11*(4), 227 (1953).

98. A. Veckenstedt, I. Beladi, and I. Mucsi, *Arch. Virol., 57,* 255
     (1978).

99. A. Wacker and H. G. Eilmes, *Arzneim. Forsch., 28,* 347 (1978).

100. M. A. Apple and L. M. Haskell, *Physiol. Chem. Phys., 3,* 307
     (1971).

101. T. S. Papas and M. P. Schafer, *Ann. N.Y. Acad. Sci., 284,* 566
     (1977).

102. M. A. Apple, L. Osofsky, W. Levinson, J. Paganelli, and E.
     Wildenrath, *Clin. Res., 20,* 562 (1972).

103. V. S. Sethi, *Ann. N.Y. Acad. Sci., 284,* 508 (1977).

104. M. Takeshita, S. B. Horowitz, and A. P. Grollman, *Virology,
     60,* 455 (1974); *Ann. N.Y. Acad. Sci., 284,* 367 (1977).

105. H. Umezawa and T. Takita, *Struct. Bonding, 40,* 73 (1980).

106. D. J. Robinson and J. H. Raschke, *J. Gen. Virol., 34,* 547
     (1977).

107. J. N. Brady, V. D. Winston, and R. A. Consigli, *J. Virol., 23,*
     717 (1977); J. Cohen, J. Laporte, A. Charpilienne, and R.
     Scherrer, *Arch. Virol., 60,* 177 (1979).

108.   A. C. Durham, *Nature*, *267*, 375 (1977); A. C. Durham and D. A. Hendry, *Virology*, *77*, 510 (1977).

109.   M. E. Rafelson, M. Schneir, and V. W. Wilson, *Arch. Biochem. Biophys.*, *103*, 424 (1963).

110.   P. P. K. Ho, C. P. Walters, F. Streightoff, L. A. Baker, and D. C. DeLong, *Antimicrob. Agents Chemother.*, 636 (1967).

111.   D. L. Swallow, *Prog. Drug Res.*, *22*, 26 (1978).

112.   R. Klöcking, K. D. Thiel, P. Wurtzler, B. Helbig, and P. Drabke, *Pharmazie*, *33*, 539 (1978); R. Klöcking, K. D. Thiel, T. Blumöhr, P. Wurtzler, M. Sprössig, and F. S·hiller, *Pharmazie*, *34*, 292, 293 (1979).

113.   D. H. Petering, *J. Med. Chem.*, *22*, 1218 (1979).

114.   V. E. Rahman, M. W. Rosenthal, and E. A. Cerny, *Science*, *180*, 300 (1973).

115.   V. E. Rahman and B. J. Wright, *J. Cell. Biol.*, *65*, 112 (1975).

116.   M. M. Jonah, E. A. Cerny, and V. E. Rahman, *Biochim. Biophys. Acta*, *401*, 336 (1975).

117.   G. Gregoriadis, E. D. Neerunjun, and R. Hunt, *Life Sci.*, *21*, 357 (1977).

118.   E. D. Neerunjun and G. Gregoriadis, *Biochem. Soc. Trans.*, *2*, 868 (1974).

119.   V. E. Rahman, M. W. Rosenthal, and E. S. Moretti, *J. Lab. Clin. Med.*, *83*, 640 (1974).

120.   J. S. Oxford and D. D. Perrin, *Ann. N.Y. Acad. Sci.*, *284*, 613 (1977).

121.   G. M. Ihler, R. H. Glew, and F. W. Schnure, *Proc. Nat. Acad. Sci. U.S.*, *70*, 2662 (1973).

122.   D. A. Tyrrell and B. E. Ryman, *Biochem. Soc. Trans.*, *4*, 677 (1976).

123.   D. A. Tyrrell, T. D. Heath, C. M. Colley, and B. E. Ryman, *Biochim. Biophys. Acta*, *457*, 260 (1976).

124.   B. E. Ryman and D. A. Tyrrell, *Essays Biochem.*, *16*, 49 (1980).

125.   Y. J. Gordon, Y. Asher, and Y. Becker, *Antimicrob. Agents Chemother.*, *8*, 377 (1975).

126.   J. Shlomai, Y. Asher, Y. J. Gordon, U. Obshevsky, and Y. Becker, *Virology*, *66*, 330 (1975).

127.   A. de Roetth, *Amer. J. Ophthalmol.*, *56*, 729 (1963).

128.   B. D. Korant, J. C. Kauer, and B. E. Butterworth, *Nature*, *248*, 588 (1974).

129.   J. Polatnick and H. L. Bachrach, *Antimicrob. Agents Chemother.*, *13*, 731 (1978).

130. B. E. Butterworth and B. D. Korant, *J. Virol., 14,* 282 (1974).

131. M. Bracha and M. Schlesinger, *Virology, 72,* 272 (1976).

132. B. E. Butterworth, R. R. Grunert, B. D. Korant, K. Lonberg-Holm, and F. H. Yin, *Arch. Virol., 51,* 169 (1976).

133. W. Levinson, A. Faras, B. Woodson, J. Jackson, and J. M. Bishop, *Proc. Nat. Acad. Sci. U.S., 70,* 164 (1973).

134. D. J. Bauer, *Brit. J. Exp. Path., 39,* 480 (1958).

Chapter 8

COMPLEXES OF HALLUCINOGENIC DRUGS

Wolfram Hänsel
Institute for Pharmacy and Food Chemistry
University of Würzburg
Würzburg, Federal Republic of Germany

1. INTRODUCTION

Hallucinogenic drugs are substances that have fundamental effects
upon the sensory perception and the mind, above all colorful optical
hallucinations [1], less often illusions of the other senses. Hallucinogenic plant substances had been known to the old cultures for
hundreds of years and most of them were used in religious and magic
ceremonies. It was in the early 1950s when juvenile subcultures
started taking them to widen the range of consciousness that interest

in these drugs gained publicity. The scientific work began after
the isolation of mescaline by Heffter in 1896 and received new and
strong impulses by the incidental discovery of the extremely strong
hallucination-producing effect of LSD by Hofmann in 1943. Occurrence, isolation, structure, activity, analyses, and synthesis methods as well as potential mechanisms of action have been described
in the literature [2-11].

Despite intense research work conducted in particular in the
last 20 years, the mechanism of action of these fascinating drugs
has not yet been clarified. With the bioinorganic chemistry becoming established and the knowledge about the significance of metal
ions for the neuronal functions continually increasing, it was necessary to find out whether, and if so, to what extent transition metals
contribute to the hallucinogenic effect [12-22].

2. THE CHEMISTRY OF THE HALLUCINOGENIC DRUGS

Today we know far more than 100 hallucination-producing drugs. It
must, however, be borne in mind that this high number is due to sometimes very slight modifications in the substitution of a small number
of fundamental structures. In view of the observed effect, the
potential mode of action, and the structural similarity to certain
neurotransmitters, it is recommended to divide hallucinogenic drugs
into psychotomimetics and deliriants [2]. In addition to these
there is a small number of drugs that cannot be unambiguously
classified.

2.1. Psychotomimetics

This group includes the greatest number of members. The effect of
these drugs has often been compared with schizophrenic conditions.
However, such a comparison applies only to a certain extent. All
these drugs are arylalkyl amines or contain the basic structure of

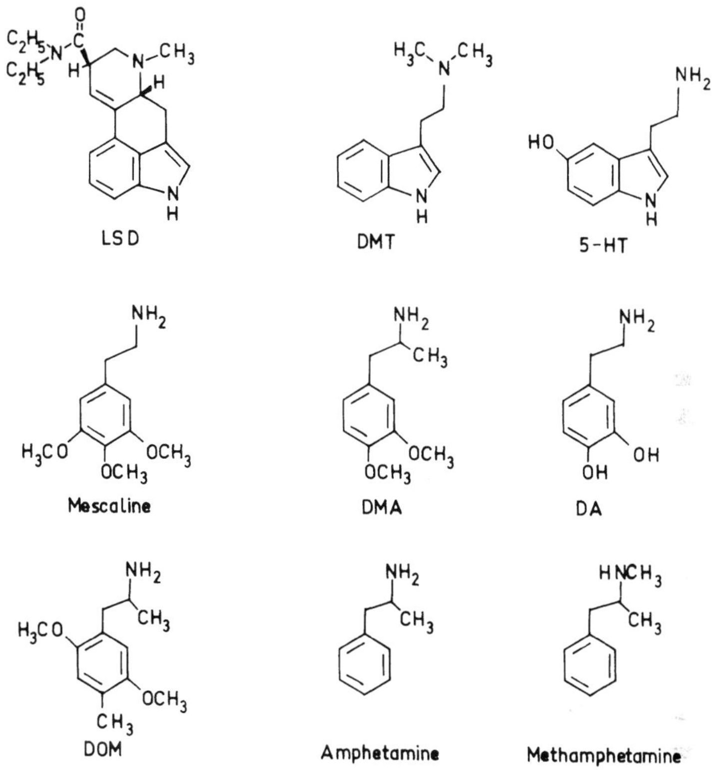

FIG. 1. Typical psychotomimetics and related compounds.

these compounds. The most important compounds are derivatives of lysergic acid, tryptamine, phenylethylamine, and phenylisopropylamine. The structural similarity to the neurotransmitters dopamine, norepinephrine, and serotonin is evident (Fig. 1).

2.2. Deliriants

The effect of this group of hallucinogenic drugs differs dramatically from that of psychotomimetics. They produce great confusion and disorientation in addition to hallucinations. All delirantia form a chemically very homogenous group. They are esters of sterically

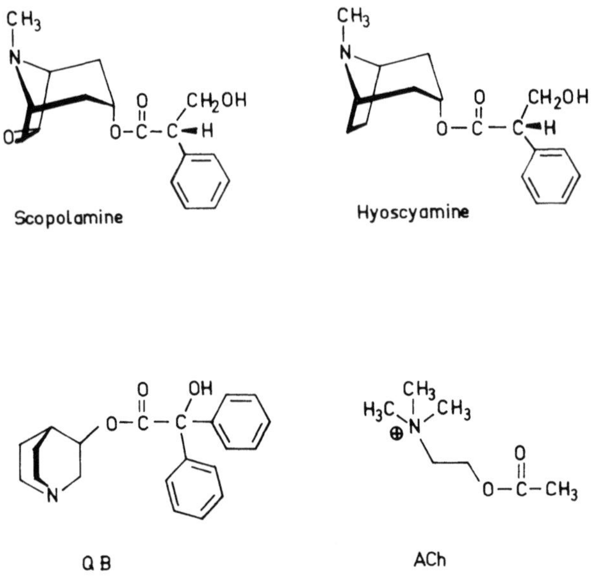

FIG. 2. Typical deliriants and acetylcholine.

bulky arylaliphatic hydroxycarboxylic acids with normally cyclic
aminoalcohols. Thus they are chemically similar to the neurotransmitter acetylcholine (Fig. 2).

2.3. Other Hallucinogenic Drugs

The members of this chemically very heterogenous group of substances
cannot be subdivided into those in Secs. 2.1 or 2.2. Kind and
strength of their effects vary widely. The most important members
are the constituents of cannabis, such as THC, phencyclidine and
its derivatives, harmine derivatives, muscimol, and ibogaine. Even
some drugs that are used as medicines have been found to produce
side effects in the form of hallucinations (Fig. 3).

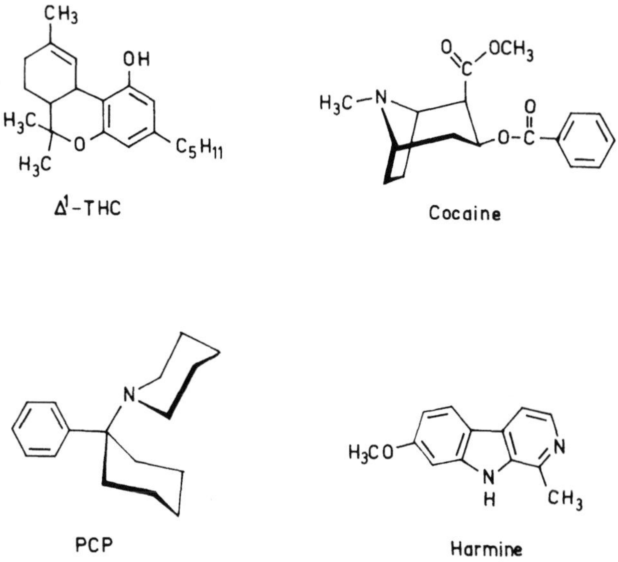

FIG. 3. Miscellaneous hallucinogens.

3. THE ROLE OF METAL IONS IN THE ACTIONS OF DRUGS

The correlation between metal ions and the actions of drugs has become increasingly interesting in recent times. Before such correlations in the case of hallucinogenic drugs are discussed in detail we should take a look at correlations in general. Roughly subdividing the possible correlations into the following groups we can distinguish between the cases when (1) the drug is a defined metal compound--the effect is due to the metal; (2) the drug is a potent chelating ligand--the effect is directly associated with the binding capacity of the metal; (3) the drug contains chelating structures in addition to other pharmacophoric groups--there is no clear correlation between effect and metal affinity; (4) the drug contains donor sites but is not capable of forming chelates--a clear correlation between the drug's capability of forming complexes and its actions has not been found.

Numerous examples, some of which have often been examined, are known of groups 1 and 2 [23-27].

With groups 3 and 4 things are more difficult. Most catecholamine neurotransmitters are chelating agents, which is obviously of essential significance for their storage in the granules in the synapses [28,29]. It has not yet been completely clarified to what extent this also applies to their effect as chemical-transmitting substances [30,31]. Unfortunately, nearly all hallucinogenic drugs are covered by group 4. In view of their structure, conspicuous correlations with biometals can thus not be expected a priori. For this reason only few investigations have been made to clarify this.

4.   COMPLEXES OF HALLUCINOGENIC AND RELATED DRUGS

Few attempts have been made so far to examine the capability of hallucinogenic drugs to form metal complexes. For this reason I should also like to report on the ability of other psychoactive substances to form complexes and on complexes of some narcotics. We must basically distinguish between complexes in which the drug is fixed to the metal atom as a genuine ligand and complex salts in which the protonated drug is the cation and a normally simple inorganic metallate is the anion. In the latter case there is no direct interaction between metal and drug. For this reason the first case is of greater bioinorganic interest.

The ability of phenylalkylamines to form complexes was the one examined in greatest detail. This group includes the psychotomimetics mescaline and dimethoxyamphetamine, the psychostimulants amphetamine and methamphetamine, as well as the model phenylethylamine itself. Misra et al. [32] reported on the preparation of mercury complexes of amphetamine, methamphetamine, and ephedrine having the general formula $HgCl_2L_2$. However, the assays carried out by infrared spectral data are controversial. Watt et al. [12-14,16,18,20,21] carefully examined the behavior of such ligands, in particular toward Cu^{2+} but also to Mn^{2+}, Fe^{2+}, Co^{2+}, and Zn^{2+} as important

TABLE 1

Solid Copper Complexes of Mescaline,
Methamphetamine, and β-Phenylethylamine

Ligand	Formula	Color	Ref.
Mescaline (MES)	$CuCl_2(MES)_2$	Purple	[12]
	$CuCl_2(MES)_2 \cdot H_2O$	Pale blue	[12]
	$CuBr_2(MES)_4$	Purple	[12]
	$Cu_4OCl_6(MES)_4$	Pea green	[12]
Methamphetamine (MA)	$CuCl_2(MA)_2$	Purple	[16]
	$CuBr_2(MA)_2$	Green	[16]
β-Phenylethyl- amine (PE)	$CuCl_2(PE)_2$	Blue	[12]
	$CuCl_2(PE)_4$	Royal blue	[12]
	$CuBr_2(PE)_3$	Turquoise	[12]
	$CuBr_2(PE)_4$	Royal blue	[12]
	$Cu_4OCl_6(PE)_4$	Pea green	[12]

biometals. A major number of well-defined copper complexes (Table 1)
could be isolated by reacting the free bases with $CuCl_2$ in an anhydrous solvent [12,16]. Complete crystal structures of these complexes
are not available. However, the infrared and electronic spectra were
carefully examined both in the solid state and in solution [12,16].
In view of these data, and especially in view of considerations concerning the symmetry, the compound of type CuX_2L_4 was suggested to
have a tetragonally distorted, octahedral structure [12]. If L is
mescaline or phenylethylamine, a clear structure of the compound of
type CuX_2L_2 cannot be derived from the spectral data. A square
planar ligand arrangement with the amine ligands in cis-position has
been suggested provided L is methamphetamine [16]. $Cu_4OCl_6L_4$ is
thought to have the structure μ4-oxo-hexa-μ-chlorotetrakis-[(phenylethylamine) copper(II)] [12]. Complexes of this structure were frequently assayed in recent times. In this cluster the ligand atoms
oxygen, halogen, and nitrogen are nearly trigonal-bipyramidal-coor-
dinated around the copper atoms [33].

If phenylethylamine hydrochlorides are used for the reaction, complex salts of the types $(LH)_2^+[MX_4]^{2-}$ and $(LH)^+[MX_3]^-$ are mostly obtained [12-14,16,18,20,21]. The only substances isolated from mescaline are tetrachlorocuprate [12,18] and tetrachloromanganate [21]. The use of methamphetamine hydrochloride yielded salts of $[CuCl_4]^{2-}$, $[CuCl_3]^-$, $[MnCl_4]^{2-}$, $[FeCl_4]^{2-}$, $[CoCl_4]^{2-}$, and $[ZnCl_4]^{2-}$ [16,20]. The assay of these salts concentrated on the characterization of the structure of the tetrachlorometallates. Some tetrachlorocuprates show thermochromism. At ambient temperature a yellow modification is obtained which is reversibly transformed into a green low-temperature modification upon cooling. Determinations of the crystal structure [13,18] have shown that in the green modification the chlorine atoms are coordinated in an almost square planar arrangement and in the yellow modification in a flattened tetrahedron.

All drugs of the phenylalkylamine type from which metal complexes could be isolated have a primary aliphatic amino function. From drugs with a secondary or tertiary amino function no complexes could be isolated, which was shown by examinations of the psychotomimetic DMT and the reference substance gramine [19]. Despite the numerous methods used, it was not possible to obtain complexes with Mn^{2+}, Fe^{2+}, Co^{2+}, Ni^{2+}, Cu^{2+}, or Zn^{2+} as solids, which can be explained by the reduced basicity of the nitrogen or by steric effects [19].

The same tendency has shown in nuclear magnetic resonance studies with 2,5-dimethoxy- and 3,4-dimethoxyamphetamine and its N-methyl and N,N-dimethyl derivatives with the shift reagent $Eu(fod)_3$ [34]. Whereas in primary amines complexes are predominantly formed at the nitrogen atom, the tertiary amines only show a weak correlation between the Eu atom and the N atom. In the latter case a likewise weak correlation with the methoxy groups can be discovered. The ability of DMT and gramine to form complexes with copper ions is indicated by studies in dioxane-water solution [19]. The formation of complexes by copper ions with cocaine and N-methylpiperidine, as a model, could also be observed in the same solvent [17]. However, in spite of the various methods used, solid complexes with Mn^{2+}, Fe^{2+}, Co^{2+}, Ni^{2+}, Cu^{2+}, and Zn^{2+} could be obtained neither with cocaine nor with N-methylpiperidine [17].

The same is true for THC [15]. This drug was examined as a ligand with oxygen as potential donor atom. Although the equilibrium with Fe^{2+}, Co^{2+}, and Cu^{2+} in tertiary butanol suggests that THC coordinated to these metals, the correlations possible are multifarious depending on the pH value, which renders a simple interpretation of the data obtained rather difficult [15].

Metal complexes of other hallucinogenic drugs, especially those of the deliriant type, have hardly been commented on. The analytic literature contains numerous references to the formation of the most various metallates of protonated drugs [35,36], mainly tetrajodobismutates, reineckates, and tetrathiocyanatocobaltates. These complexes are generally used for the qualitative and quantitative detection of nitrogen-containing drugs. Little is known about the structure of such complexes. In view of the reaction conditions we may, however, expect that the drug is present as a cation and that there is no direct interaction between the drug and the metal. The tetrathiocyanatometallates of harmine which have the composition (harmine H)$_2^+$[M(SCN)$_4$]$^{2-}$ (M = Co^{2+}, Zn^{2+}, Hg^{2+}) could be isolated in crystalline form [37,38]. The composition of the corresponding cobaltates of phenothiazines is much more complex [39]. Phenothiazines are antagonists of most psychotomimetics and thus of neurochemical importance in the present case. It is interesting to know that specific 1,4-benzodiazepines, such as diazepam, which are also used as antidotes in case of intoxication by psychotomimetics, form well-defined complexes having the composition $M_2L_3X_4$ (M = Co^{2+}, Ni^{2+}; X = Cl^-, Br^-, J^-; L = diazepam) and ML_2X_2 (M = Cu^{2+}; X = Cl^-, Br^-) which can be isolated [40]. These complexes were well characterized by infrared and electronic spectra as well as by magnetic measurements [40]. It could be shown that diazepam is bound to the metal via nitrogen.

Finally, I should like to draw the reader's attention to complexes of CNS-active barbiturates. Barbiturates are occasionally

misused as narcotics. It has been known for a long time that they
form purple complexes with cobalt(II) ions in the presence of bases
(so-called Zwikker reaction) [35,41]. They also form stable complexes with silver and mercury ions, which is useful for their quantitative determination [35]. Recently it was possible to find out
the structures of the cobalt and zinc complexes of barbital (with
imidazol as base) [42] and the copper complex of barbital (with
pyridine as base) [43] by an analysis of their crystal structure.
The composition of these complexes is $M(barbital)_2B_2$. The coordination of the cobalt and zinc complexes is tetrahedral and that of the
copper complex square-planar. Barbital is bound to the complex at a
nitrogen atom.

5. CONCLUSIONS

Although the mechanism of action of hallucinogenic drugs has not yet
been completely clarified, all psychotomimetics are generally assumed to act in the same way as presynaptic serotonin and dopamine
mimetics in certain regions of the midbrain [2,44,45]. Recently
they have even been thought to interact with opiate receptors [46].
Deliriants, on the other hand, probably act as central anticholiner-
gics [3,10]. If these hallucinogenic drugs interfere with neurotransmitters, it seems possible that metal ions may contribute to
the hallucinogenic effect.

However, all data available so far show that hallucinogenic
drugs do not have a strong tendency to form metal complexes under
physiological conditions. This suggests that, apart from their
general significance for the neural functions, metal ions do not
directly contribute to the hallucinogenic activity. However, considering their capability to form complexes, metal ions might well
contribute to the effect of barbiturates. Although quantum mechani-
cal calculations [47] concerning the correlation between barbiturate
and receptor could be made without considering metal ions, this complex of problems is worth being investigated in further detail.

ABBREVIATIONS

ACh	Acetylcholine
DA	Dopamine
DMA	Dimethoxyamphetamine
DMT	Dimethyltryptamine
DOM	Dimethoxymethylamphetamine
5-HT	5-Hydroxytryptamine (serotonin)
LSD	Lysergic acid diethylamide
MA	Methamphetamine
MES	Mescaline
PCP	Phencyclidine
PE	β-Phenylethylamine
QB	3-Quinuclidinyl benzilate
THC	Tetrahydrocannabinol

REFERENCES

1. R. K. Siegel, *Sci. Am.*, *237*, 132 (1977).

2. W. Hänsel, *Chem. Unserer Zeit*, *13*, 147 (1979).

3. R. A. Heacock, *Progr. Med. Chem.*, *11*, 92 (1975).

4. A. R. Patel, *Prog. Drug Res.*, *11*, 11 (1968).

5. K. A. Nieforth, *J. Pharm. Sci.*, *60*, 655 (1971).

6. S. Cohen, *Progr. Drug Res.*, *15*, 68 (1971).

7. T. D. Inch and R. W. Brimblecombe, *Int. Rev. Neurobiol.*, *16*, 67 (1974).

8. L. G. Abood, in *Drugs Affecting the Central Nervous System* (A. Burger, ed.), Marcel Dekker, New York, 1968, pp. 127 ff.

9. R. W. Brimblecombe and R. M. Pinder, *Hallucinogenic Agents*, Wright-Scientechnica, Bristol, 1975.

10. R. C. Stillman and R. E. Willette, eds., *The Psychopharmacology of Hallucinogens*, Pergamon Press, New York, 1978.

11. A. Hofmann, *LSD: Mein Sorgenkind*, Klett-Cotta, Stuttgart, 1979.

12. G. W. Watt and M. T. Durney, *Bioinorg. Chem.*, *3*, 315 (1974).

13. R. L. Harlow, W. J. Wells III, G. W. Watt, and S. H. Simonsen, *Inorg. Chem.*, *13*, 2106 (1974).

14. R. L. Harlow, W. J. Wells III, G. W. Watt, and S. H. Simonsen, *Inorg. Chem., 13,* 2860 (1974).

15. G. W. Watt and J. R. Paxson, *J. Inorg. Nucl. Chem., 38,* 627 (1976).

16. G. W. Watt and W. J. Wells III, *J. Inorg. Nucl. Chem., 38,* 921 (1976).

17. H. C. Nelson and G. W. Watt, *J. Inorg. Nucl. Chem., 41,* 99 (1979).

18. H. J. Buser and G. W. Watt, *Acta Chrystallogr. Sect. B., 35,* 91 (1979).

19. H. C. Nelson and G. W. Watt, *J. Inorg. Nucl. Chem., 41,* 406 (1979).

20. G. W. Watt and D. S. Burz, *J. Inorg. Nucl. Chem., 41,* 405 (1979).

21. S. E. Hoppe, H. C. Nelson, J. B. Longenecker, and G. W. Watt, *J. Inorg. Nucl. Chem., 41,* 1507 (1979).

22. H. C. Nelson, S. H. Simonsen, and G. W. Watt, *Chem. Commun.,* 632 (1979).

23. H. Sigel, ed., *Metal Ions in Biological Systems,* Vol. 11, Marcel Dekker, New York, 1980.

24. D. D. Perrin, in *Topics in Current Chemistry,* Vol. 64 (F. Boschke, ed.), Springer, Heidelberg, 1976, pp. 181 ff.

25. R. D. Gillard, in *New Trends in Bio-inorganic Chemistry* (R. J. P. Williams and J. R. R. F. Da Silva, eds.), Academic Press, London, 1978, pp. 355 ff.

26. R. Haller and W. Hänsel, *Pharm. Zeitung, 118,* 786 (1973); *Chem. Abstr., 79,* 101169 (1973).

27. W. Hänsel and R. Haller, *Pharm. Zeitung, 118,* 987 (1973); *Chem. Abstr., 80,* 43788 (1974).

28. K. S. Rajan, R. W. Colburn, and J. M. Davis, in *Metal Ions in Biological Systems,* Vol. 6 (H. Sigel, ed.), Marcel Dekker, New York, 1976, pp. 291 ff.

29. A. Gergely and T. Kiss, in *Metal Ions in Biological Systems,* Vol. 9 (H. Sigel, ed.), Marcel Dekker, New York, 1979, pp. 143 ff.

30. M. A. Bronstrom, C. O. Bronstrom, B. Mcl. Breckenridge, and D. J. Wolff, *Advanc. Cyclic Nucleotide Res., 9,* 85 (1978).

31. S. H. Snyder and R. R. Goodman, *J. Neurochem., 35,* 5 (1980).

32. C. H. Misra, S. S. Parmar, and S. N. Shukla, *Can. J. Chem., 46,* 2685 (1967).

33. F. A. Cotton and G. Wilkinson, *Advanced Inorganic Chemistry,* 4th ed., Wiley, New York, 1980, p. 817.

34. R. V. Smith, P. W. Erhardt, D. B. Rusterholz, and C. F. Barf-
     knecht, *J. Pharm. Sci.*, *65*, 412 (1976).

35. S. Ebel, *Handbuch der Arzneimittelanalytik*, Verlag Chemie,
     Weinheim, 1977.

36. R. Kakač and Z. J. Vejdelek, *Handbuch der photometrischen
     Analyse organischer Verbindungen*, Verlag Chemie, Weinheim, 1974.

37. V. I. Orlova and A. I. Portnov, *Nekotorye Voprosy Farmatsii,
     Sbornik Nauch. Trudov Vyssh. Farm. Ucheb. Zavedenii Ukr. SSR*,
     42 (1956); *Chem. Abstr.*, *53*, 6538d (1959).

38. V. I. Orlova and A. I. Portnov, *Issledovaniya v Oblasti Farm.
     Zaporozh. Gosudarst. Farm. Inst.*, 29 (1959); *Chem. Abstr.*, *55*,
     3007i (1961).

39. J. Lagubeau and P. Mesnard, *Bull. Soc. Chim. Fr.*, 2815 (1965).

40. C. Preti and G. Tosi, *J. Inorg. Nucl. Chem.*, *41*, 263 (1979).

41. R. Abu-Eittah and A. Osman, *J. Inorg. Nucl. Chem.*, *41*, 1079
     (1979).

42. B. C. Wang and B. M. Craven, *Chem. Commun.*, 290 (1971).

43. M. R. Caira, G. V. Fazakerley, P. W. Linder, and L. R. Nassim-
     beni, *Inorg. Nucl. Chem. Lett.*, *9*, 1101 (1973).

44. R. S. Sloviter, E. G. Drust, B. P. Damiano, and J. D. Connor,
     J. Pharmacol. Exp. Ther., *214*, 231 (1980).

45. H. J. Haigler and K. Aghajanian, *Fed. Proc. Fed. Amer. Soc.
     Exp. Biol.*, *36*, 2159 (1977).

46. U. Braun, A. T. Shulgin, and G. Braun, *Arzneim. Forsch.*, *30*,
     825 (1980).

47. H.-D. Höltje, *Arch. Pharm. Weinheim Ger.*, *310*, 650 (1977).

Chapter 9

LITHIUM IN PSYCHIATRY

Nicholas J. Birch[*]
Department of Biochemistry
The University of Leeds
Leeds, England

1. INTRODUCTION

Of the population of the United Kingdom, 1 in 2 000 receives lithium
for the treatment of manic-depressive psychoses, some 25 000

---

[*]*Present affiliation:* Department of Biological Sciences, The Polytechnic, Wolverhampton, England.

patients [1]. Lithium has been used for 30 years, but there is
still no agreement on its mode of action. However, it is pharmacologically intriguing since it is the lightest solid element and its
chemistry is simple, and it might therefore be argued that a demonstration of its actions at a molecular level might have fundamental
significance in the understanding of biochemical processes and the
interaction with them of inorganic species. For this reason I believe that an interdisciplinary approach is required, and the intention of this review is to present some aspects of the clinical, pharmacological, and biochemical background, together with more detailed
discussion of particular areas of current research activity. Lithium
has been the subject of a number of recent monographs and symposia;
these are listed in Table 1.

TABLE 1

Major Literature Sources on Lithium in Medicine

Date	First author	Pages	Clinical	Pharmaco-logical	Bio-chemical	Ref.
Edited monographs						
1973	Gershon, S.	358	*	*	*	2
1975	Johnson, F. N.	569	*	*	*	3
1978	Johnson, F. N.	459	*	*	*	4
1980	Johnson, F. N.	453	*			5
Symposia						
Psychiatric						
1976	Bunney, W. E.	96		*	*	6
1979	Cooper, T. B.	984	*	*	*	7
Granulopoiesis						
1980	Rossof, A. H.	475	*			8
Complete cumulative bibliography						
Covers 1818-1980	Schou, M.		*	*	*	9-16

Key: * = major content.

Lithium was discovered by Arfwedson in 1818 during analysis of the mineral petalite [17]. Berzelius, in whose laboratory the work was carried out, suggested the name *lithion*. Garrod (1859) [18] introduced lithium into medical practice for the treatment of gout since lithium urate is the most soluble salt of uric acid and would thus be uricosuric. The element was later detected in a number of spa waters in Europe, and this led to the common belief that it contributed to the efficacy of spa treatments which were fashionable at that time. Thilenius (1882) [19] reported concentrations of 1 mmol/liter lithium in spa waters. A more specific indication for lithium therapy was revealed in the early years of this century when it was noted that of all the bromides, lithium bromide was the most effective as a sedative. The known uses of lithium at that stage were reviewed by Squire (1916) [20]. However, the hypnotic properties of the bromide ion itself were demonstrated and the possibility of an action of the lithium ion itself was not considered.

These and other temporarily fashionable uses of lithium continued until the late 1940s, and these treatments shared the characteristic that lithium was used in relatively small doses prescribed by a physician or alternatively in dilute solutions. Daniels (1914) [21] recommended between 1 and 2 g of lithium citrate daily for the treatment of gout, though Cleaveland (1913) [22] had reported that between 4 and 8 g of lithium chloride per day gave transitory muscular weakness and psychiatric symptoms. An early pharmacology textbook [23] listed the dose of lithium carbonate as 0.7-2 g/day.

During the 1930s it became clear that patients with cardiac failure or hypertension benefited by severe restriction in intake of sodium chloride. The main disadvantage of this course of action was the decreased palatability of food cooked with the exclusion of salt, and in order to make the salt-restricted diet acceptable efforts were made to find a salt substitute with which to flavor the food. An aqueous solution of lithium chloride with citric acid and traces of potassium iodide appeared to be an acceptable taste substitute [24], and a number of trials were carried out when it was found that the average consumption of lithium chloride was about 0.5 g/day. After a number of such trials in the United States the salt substitute was released for general use. However, unlike the situation in the clinical trials a number of patients suffered lithium toxicity as a result of ad libitum use of the salt substitute. Several deaths were reported following intakes of 4 to 6 g/day [15]. Talbott (1950) [24] describes the precipitate action of the drug-regulating authorities and suggests that an unnecessary panic was induced in several thousands of patients taking salt substitutes and bewilderment in their physicians who had not been properly informed of the decisions. These panic measures, however, had a lasting effect on the subsequent history of lithium. It is now known that cardiovascular diseases, and in particular low salt intake, are contraindications for lithium therapy, but more important is the very strict control of intake.

Meanwhile, in Australia, Cade had completed his initial studies of lithium in the treatment of mania and his results were published in ignorance of the American experience [26,27].

Cade's report of the successful use of lithium in mania coincided almost exactly with the publication of reports by Corcoran et al. [28] and Hanlon et al. [25], and the auguries for its successful adoption by the medical community were not good. Not only was there intense mistrust as a result of the American reports, but Cade's short report [26] appeared in a journal with very little circulation outside his native Australia and he was an unknown psychiatrist working in an obscure, small mental hospital. He was fortunate, therefore, that the paper came to the attention of a small number of psychiatrists in Europe, in particular Mogens Schou who was prepared to try the drug in a controlled manner in a small number of patients. It should also be noted that at that stage, the early 1950s, there were very few drugs of any sort with which to treat disturbed psychiatric patients, and its success was therefore all the more remarkable to those psychiatrists who were courageous enough to try it.

The subsequent history of the controversies and the difficulties of acceptance have been well recorded by many authors and is the subject of reviews in the volumes listed in Table 1. Lithium is no longer used in the treatment of acute manic excitement since modern phenothiazine tranquilizers have a more rapid action and are less toxic. The major place for lithium in current practice is in the preventative, or prophylactic, treatment of recurrent manic-depressive episodes. The major credit for rehabilitation of lithium must go to Schou who, with his colleague Baastrup, laid the foundation for its modern use.

2. LITHIUM IN PSYCHIATRY

2.1. Manic-Depressive Psychoses

Manic-depressive psychoses (recurrent affective disorders) are disabling and distressing diseases both for the patient and for relatives. These diseases occur more frequently in women than in men and in particular after the menopause. This is shown in Fig. 1,

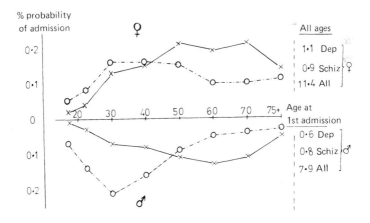

FIG. 1. Percentage probability of admission to a psychiatric hospital for the *first* time, at various ages, for treatment of (a) depressive psychoses (including manic-depressive psychoses) (x——x); (b) schizophrenias (o----o). Cumulative probability of admission to psychiatric hospital for depressive, schizophrenic, and all psychiatric disease, at some stage during life, is also given. (Redrawn, with the permission of The Comptroller, Her Majesty's Stationery Office, from data published in D.H.S.S. Statistical Report No. 20; In-Patient Statistics from the Mental Health Survey, 1975 H.M.S.O., London, 1978.)

which is compiled from data in a recent mental health survey of
Great Britain. Figure 1 also draws comparison with the incidence of
schizophrenia, and it is clear that not only does schizophrenia have
a much earlier age of onset, but there is a much higher frequency of
male patients. There is a very clear clinical difference between
these diseases. It should be understood, however, that psychiatric
classification is only in terms of symptoms presented by the patient,
and patients displaying these symptoms are categorized accordingly.
This does not imply that each is a single nosological entity, and
each category may contain a number of different disease processes
producing the same symptoms. Psychoses may be divided into three
groups: (1) toxic psychoses, when a known toxic agent has been absorbed by the body, for instance heavy metals; (2) organic psychoses,
where there is brain damage as a result of injury or degenerative
disease; and (3) the functional psychoses, schizophrenia and the affective disorders, both of unknown origin.

The affective disorders show altered mood (affect), the older
term *manic-depressive psychoses* generally being reserved for those
patients in whom a cyclical course of the disease is apparent.

The manic phase is the most immediately obvious because the
patient shows hyperactivity, increased verbal activity accompanied
by flights of ideas, which may be grandiose. The behavior and dress
of the patient may be bizarre. Other typical characteristics are interfering, argumentative, and belligerent behavior, sometimes excessive drinking, and a tendency to be profligate with their own posses-
sions and less than scrupulous in the disposal of other peoples'.
Above all, though, they are extremely happy, and this is a phase of
the disease which is enjoyed by many patients and can be very profit-
able in the case of creative writers, artists, and performers whose
output increases enormously [29]. Mania is, however, dreaded by the
relatives and close associates of the patient since the excessive
level of activity and pressure of speech are extremely wearing, and
social problems may arise as a result of excessive spending, alcoholic intake, petty pilfering, and gratuitous aggression.

Depression, on the other hand, is the most distressing part of the disease for the patient. Not only is the thought content self-critical and morbidly preoccupied, but ideas come slowly and the inability to summon up the energy to act, even in the case of a simple task, is itself traumatic and frequently misunderstood by those around. This leads to a further twist in the self-recrimination cycle. Suicidal thoughts often predominate, but the patient is unable to act upon them and in this respect the recovery period from the depression is the most dangerous.

In many patients a single episode of depression occurs as a result of, for instance, the death of a loved one or severe financial difficulties. Such an event may not recur. Nevertheless, in a large number of patients depression or mania may occur with no premonitory event, and these episodes may occur at gradually increasing intervals. In some patients a particularly regular cycle is established which may be very persistent.

One aspect of the disease process which is particularly fascinating, and which has so far evaded all answers, is the cyclical, or periodic, nature of the disease course. There is a very small number of patients who show extremely rapid and regular fluctuations in mood. We have studied one such patient in Leeds, and an example of her mood chart over a short period is given in Fig. 2. For several years the characteristic pattern of this lady was to change from mania into depression overnight, to spend the whole of the subsequent day depressed, to awake depressed on the second day, and switch into mania following a short nap after lunch. The manic mood then continued for the rest of the second day and the whole of the third day, at which stage she again returned to depression upon awakening the following morning. This was not an exact cycle and a number of abnormalities occurred. However, over a long period of time the overall cycle was maintained. In another subject, described by Jenner and his colleagues [30], an extremely regular and persistent cycle occurred. The patient, in this case a man, for a period of 11 years had a consistent cycle in which one day of mania was followed by a day of depression. During the 7 years in which

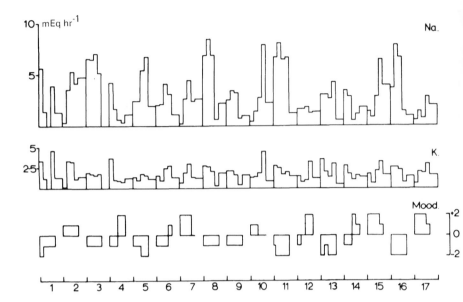

FIG. 2. Metabolic balance study on a 42-year-old woman with short-cycle periodic affective disturbance. Mood is shown from nursing records using an individually designed rating scale which was applied every 4 hr. Mean scores for each 4-hr period from five separate indices of mood were aggregated and a global rating simplified to the range (+2, mania, to -2, depression) shown by the vertical bars. Sleep is indicated by absence of mood. Also shown are the excretory rates of sodium and potassium throughout this period of 17 days. (From M. N. E. Allsopp, N. J. Birch, and R. P. Hullin, unpublished data.)

this cycle was accurately recorded only 9 defects in cycle occurred, and in fact the periodicity was so predictable that the wedding of the patient's daughter was arranged so that it would occur on a day on which Father would be in a depressed state--then he would be less interfering and disruptive.

In both of these short-cycle cases lithium treatment was instituted. In the woman with the 72-hr cycle lithium was relatively ineffective and did not stabilize her mood. The length of cycle was altered, but not consistently. The 48-hr periodic psychotic man, however, provides a most spectacular example of lithium therapy [31]. When he was first given the drug his cycles became gradually attenuated, and he was eventually restored to apparent normality. At this

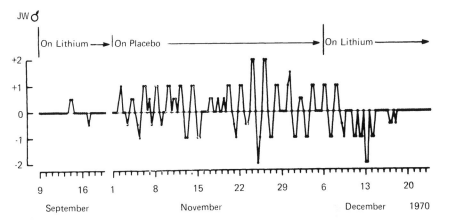

FIG. 3. Daily mood chart of a 57-year-old man with 48-hr periodic
affective disturbance who had been stabilized on lithium for 3 years
prior to discontinuation under clinical surveillance. At the end of
the period of placebo treatment his lithium treatment was recommenced.
(From Ref. 31, by kind permission of the copyright holder.)

stage he was able to find employment, impossible for the previous 11
years, and worked successfully until his retirement some years later.
His lithium treatment was discontinued briefly under careful clinical
control, and it was clearly demonstrated that the underlying cycle
was still present since his moods began to fluctuate at its original
rate, and this was once again abolished by the administration of
lithium (Fig. 3).

Such very-short-cycle patients are extremely rare, but they do
emphasize the importance of discovering the mechanism by which the
mood change is triggered, and it is clear that lithium must have one
of its actions on this trigger mechanism, though whether this is its
major action remains to be seen. A large number of processes in biological systems are regulated in a cyclical manner, and these cycles
vary from those which take a year or more to complete, such as breed-
ing cycles, down to the very rapid cycles encountered in oscillating
enzyme systems [32]. One of the most persistent and pervasive of
these cycles is the diurnal rhythm. Defects in mood regulation
could be envisaged in terms of desynchronization of two or more of

such regulatory cycles. "Beat" phenomena could readily be antici-
pated which might provide adequate stimulus for a switch. Jenner et
al. [30] discussed in detail these problems and described a study of
their 48-hr psychotic patient during a period in which his "day" was
artificially adjusted to 22 hr real time. Other aspects of the
"switch" mechanism have been reviewed [33,34].

2.2. Clinical Use of Lithium

The majority of patients who are treated successfully with lithium
are, however, not of such short cycle as those described above. Indeed, although they suffer recurrent attacks, these are not necessarily predictable. Lithium is not usually used in the acute treatment of mania, and its major use is therefore as a preventative
treatment and as such must be administered regularly for a long
period. Various criteria have therefore been suggested for the psychiatrist to be able to make the decision whether it is justifiable
to embark upon long-term treatment. As a rough guide, lithium might
be administered to a patient who has had to have two admissions to
hospital for either mania or depression during the preceding 2 years.
The criteria for selection of patients have been discussed in great
detail [5,35].

Lithium is used mainly in the treatment of bipolar periodic
affective disorder, that is, when both manic and depressive phases
have been seen in the patient's history. Evidence is now appearing
that lithium is also effective in unipolar, depressive affective disorder [36,37]. Other psychiatric indications have been reviewed by
Schou [38,39], who reports that it has been suggested for use in recurrent schizoaffective disorder, emotional instability in children
and adolescents, pathological impulsive aggression, premenstrual
tension syndrome, alcoholism, drug addiction, and affective disorders
resulting from organic brain syndromes. Nonpsychiatric uses have
also been reviewed [38,40] and the following uses have been reported:

1. Movement disorders (Huntington's chorea, tardive dyskin-
    esia, L-dopa-induced hyperkinesis in Parkinsonism, spas-
    modic torticollis)
2. Thyroid diseases (hyperthyroidism, thyroid cancer)
3. Inappropriate secretion of antidiuretic hormone
4. Granulocytopenia
5. A miscellaneous group of diseases including migraine and
    cluster headache, periodic hypokalaemic paralysis,
    cyclical vomiting, and Menière's disease

The use of lithium in the treatment of granulocytopenia has
developed rapidly, and a recent book [8], the first on this topic,
brings together much of the basic information. This work is beyond
the purview of the present psychiatric review but is a major source
of further information on lithium action.

2.2.1. Practical Aspects of Lithium Prophylaxis

By nature, lithium prophylaxis is a long-term process and may be
likened to the initiation of insulin therapy in diabetic patients.
The therapeutic index of lithium is low, and thus patients must be
clinically assessed to reveal any preexisting medical conditions
which might affect the regulation of plasma lithium, and in particular its excretion by the kidney. Regular monitoring and recording
of plasma lithium concentrations are essential, and monitoring of
thyroid and renal function is strongly advised. The objective of
the frequent monitoring is to try to maintain the plasma lithium
within set limits; originally these were between 0.6 and 1.2 mM/liter,
though in the light of recent experience we would recommend a range
of 0.4 to 1 mM/liter [41-43]. Toxic symptoms are seen often if the
plasma lithium rises above 2mM/liter. In most patients a dose of
between 1 and 2 g of lithium carbonate each day will produce plasma
lithium in the recommended range after the "steady state" has been
attained.

There is considerable interindividual variation in dose depending on the volume of distribution, which is related to body size,

and on the state of kidney function. Elderly patients, whose renal function is gradually reduced, may require doses as low as 400 mg of lithium carbonate per day to maintain the therapeutic level, while younger patients may require much higher doses [44,45]. The technique of initiation of therapy is therefore either to perform one of the predictive lithium clearance tests (e.g., those of Schou [46] or Cooper et al. [47,48]) or alternatively to institute a dosage regime which starts at a low dose and is increased weekly by very small increments with plasma lithium determination immediately prior to each change of dose. Because of the monitoring and supervision required during the early stages it is often considered desirable to start lithium on an in-patient basis though this is not essential provided the patient has a reliable and capable relative or companion who is able to ensure that the tablets are taken as directed and to report any unusual or untoward signs which might indicate impending toxicity.

Blood samples are taken weekly for the first month and then the interval is gradually increased until visits for monitoring occur at about 6 to 8 weeks. In elderly patients and in patients with cardiovascular or renal disease and those who are known to be unreliable in taking medication this interval is reduced. It is vitally important that good records are kept of the progress of the treatment and of the plasma lithium concentration at each visit, and long-term records of renal function, urea, creatinine, and thyroid function should also be maintained. In this way the lowest effective dose can be maintained and the development of any long-term side effects noted before intoxication occurs.

2.3. Side Effects and Toxicity

In the early stages of lithium treatment it is very common to see minor side effects such as a slight tremor of the hand and mild intestinal upsets. However, these symptoms disappear in the majority of cases within a few weeks. In about 15% of patients hypothyroidism

is seen, and this is treated by small doses of thyroxine. The warning signs of impending intoxication are a coarse hand tremor, drowsiness, dizziness, and slurred speech together with vomiting and
diarrhea [46]. In such cases discontinuation of lithium will re-
lieve the symptoms.

However, if the warning signs are missed and frank intoxication
occurs, perhaps as a result of suicidal attempt, the main effects are
on the central nervous system [49], although kidney damage may lead
to rapid accumulation of lithium [50,51]. Hemodialysis is the best
treatment for intoxication, though it must be remembered that the
body load of lithium may be relatively resistant to removal, and following dialysis the plasma lithium may rise again dramatically and
further dialyses may be required.

Lithium poisoning occurs as a result of unnecessarily high
maintenance dosage of lithium, acute poisoning due to suicidal attempts, or reduced kidney function and hence lithium excretion. Reduced renal function may be as a result of slowly developing kidney
disease, sodium deficiency, or dehydration, the two latter being a
consequence of, for instance, unaccustomed exposure to hot climates,
starvation, or salt-wasting diuretics. One particular problem in
the management of obese female patients is that they may undertake a
slimming diet with consequent decrease of sodium intake and the
physician may be unaware of this change in habit. In patients who
are regularly monitored, however, we have found that the development
of intoxication is very rare since premonitory signs can be seen at
an early stage and the dosage of lithium reduced accordingly.

The mechanism of the development of lithium intoxication has
been studied in very great detail by Thomsen and various colleagues
[50-52]. The final stages of lithium intoxication affect mainly the
nervous system; these effects are, however, principally the result of
the rapid rise in plasma lithium due to its effects on renal function.
The main route of lithium excretion is through the kidney, and in a
normal state plasma lithium is regulated by the excretion of an
amount equivalent to the daily intake. The lithium cation is filtered at the glomerulus and this filtrate passes to the proximal tubule

where about 80% of lithium is reabsorbed together with water and
sodium. The remaining 20% of the original filtered lithium now
passes to the distal renal tubule where very little is absorbed and
most of it is excreted in the urine. This mechanism differs from
that of sodium reabsorption, where 80% is absorbed in the proximal
tubule, and the fine control of sodium excretion is achieved in the
distal tubule, where up to 95% of the remaining sodium may be reabsorbed. If there is a fall in dietary sodium, the body conserves
sodium by reabsorbing more in the distal tubule. However, should
this mechanism not be adequate a further system comes into play and
proximal tubular reabsorption of sodium is increased. In lithium-
treated patients this increased proximal sodium reabsorption is associated with an increase in lithium reabsorption. The possibility
of lithium toxicity arises because lithium inhibits the reabsorption
of sodium in the distal tubule because of its effect on response to
aldosterone. The renal response to this is to increase proximal
sodium reabsorption and hence proximal lithium absorption. The increased lithium absorption results in increased plasma lithium which
in turn exacerbates the sodium loss.

In a normal situation with a fluctuating plasma lithium following successive doses of lithium the plasma lithium concentration is
able to fall sufficiently to enable the sodium conservation mechanism
to return toward normal. The concomitant reabsorption of lithium in
the proximal tubule is therefore also reduced. However, when the
dose of lithium is increased, or when an additional stress such as
water or sodium depletion is added, the effective clearance of
lithium by the kidney is progressively decreased since the sodium
conservation mechanism is operative for a larger proportion of the
between-dose interval. Since the plasma lithium as a consequence
rises, this will further stimulate sodium retention in the proximal
tubule and ultimately a fulminating situation arises leading to
rapidly increasing plasma lithium concentrations and neurological
sequelae.

A number of diuretics, particularly the thiazides, inhibit
distal sodium reabsorption and for this reason may lead to sodium

loss. From the above discussion it is clear that such diuretics can readily precipitate lithium intoxication due to the compensation of the proximal tubule. However, it has been reported that, provided monitoring of plasma lithium is carried out very frequently, it is possible to maintain patients on lithium despite the necessity for their continued diuretic therapy [53]. A reduced lithium dose is obviously required.

Some side effects are seen in the long-term-treated patients and these are often referable to the various systems on which lithium acts. Table 2 lists the various side effects which may be seen at

TABLE 2

Side Effects of Lithium

Apparently harmless side effects seen during maintenance lithium therapy

	First few days (peak 1-2 hr after dose)	First 2 months	Long-term
Nausea	————————|		
Abdominal upset	————————|		
Loose stools	————————|		
Thirst	————————|		
Urine volume and frequency increased	————————|		
Fine tremor of hand	——————————————————————————————→		
Fatigue and lethargy		——————————————————→	
Polydipsia-polyuria		——————————————————→	
Minor ECG change		——————————————————→	
Hypothyroidism			————————→
Leukocytosis			————————→
Weight gain			————————→
Edema			————————→

TABLE 2 (Continued)

Toxicity

	Impending toxicity	Li intoxication
Vomiting and diarrhea	⟶	
Coarse hand tremor	⟶	
Sluggishness and sleepiness	⟶	Impaired consciousness⟶Coma⟶Death
Vertigo	⟶	
Lack of appetite	⟶	
Polyuria	⟶	Oliguria⟶ Anuria
Epileptiform seizures		⟶
Muscle twitches		⟶
Nystagmus		⟶

different stages of therapy. For clinical details the reader is referred to the handbook edited by Johnson [5].

3.   CHEMICAL ASPECTS OF LITHIUM

The chemistry of lithium with respect to its use in the periodic affective disorders has been reviewed by Williams [54,55] and many
aspects have been enlarged upon by contributors to the volume edited
by Bunney and Murphy [6].

Lithium is the lightest member of Group I, the alkali metals,
in the periodic table. Because of its first row position its chemistry is anomalous and may be likened, in many respects, to magnesium
and calcium, which are members of Group II. This so-called diagonal
relationship is a consequence of the relatively large increment in
charge and size as one passes along the first row of the periodic

TABLE 3

Physical and Chemical Properties of
Groups I and II Elements

	K	Na	Li	Mg	Ca
Atomic radius (Å)	2.03	1.57	1.33	1.36	1.74
Crystal radius (Å)	1.33	0.95	0.60	0.65	0.99
Hydrated radius (Å)	2.32	2.76	3.40	4.67	3.21
Polarizing power (z/r^2)	0.56	1.12	2.80	4.70	2.05

table and the characteristically large decrease in size as one passes
down each group from the first to the second member. The results of
this can be seen from the physical and chemical properties of lithium
and its close neighbors in the periodic table (Table 3). It is clear
that lithium is closer in atomic and ionic radius to magnesium than
to sodium. Its hydrated radius is between that of magnesium and calcium, as is its polarizing power.

Lithium has a much greater covalent character than do the other
alkali metals as a result of the very small size of its atom and ion
and hence a concentration of charge. The small size of the lithium
ion, however, produces steric restrictions on hydration in crystals
of its salts unless the anion is itself hydrated: its crystalline
hydrates usually contain no more than four water molecules.

In solution, the small crystal radius and high charge density
result in a large hydration sphere whose size is uncertain. This is
a result of the very small diameter of lithium in relation to the
aqueous solvent. The normal considerations of Stoke's law do not
apply and the mobility data must be corrected [56]. As a further
consequence, lithium salts do not conform to ideal solution behavior.

In common with the other alkali metal ions, the complex chemistry of lithium has been relatively neglected. This is because the
noble gas structure of the cations was considered to provide little
opportunity for complex formation. Nevertheless, Williams [54,55]
has suggested that because of its high polarizing power lithium might

cause distortion in the electron clouds of ligands, and indeed the
ion has a higher affinity for oxygen and nitrogen sites than do the
other metals of Group I. Because of this apparent nonselectivity of
lithium it was difficult to reconcile the pharmacological effects in
terms of occupation of receptor sites on effector molecules--for instance, in membranes or in enzymes. It was clear that such interactions must take place because of the known activating and inhibiting
effects of other alkali metals on various enzyme processes, and these
were interpreted in terms of ionic interactions at stereochemically
defined sites with rather selective dimensions.

Recently it has become possible to investigate more fully the
interactions of the alkali metals with biological systems because of
the recognition by Moore and Pressman [57] that a number of the
naturally occurring antibiotics selectively complexed particular
metals to increase their rate of transport across membranes. Valin-
omycin complexes potassium and transports it much more effectively
than it does sodium. By contrast, actinomycin complexes sodium in
preference to potassium.

Families of ligands with such ion-transporting properties have
been synthesized in the laboratory with a range of metal specifici-
ties [58]. The most interesting of these compounds with respect to
lithium pharmacology is the cryptand (Fig. 4) described by Lehn [59]
where n = 1, 2nd bridge m = 0, and 3rd bridge is —(CH$_2$)$_5$—. The
order of affinity of this transporting agent across membranes was
lithium:sodium:potassium = 1:1.8:0, and it was suggested that further
modification might increase the selectivity for lithium.

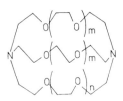

FIG. 4. Basic structure of cryptands. (Redrawn with the kind permission of J.-M. Lehn from *Struc. Bonding*, *16*, 2, 1973.)

Weak but selective interactions may occur between the alkali metals and biological molecules, and indeed we have studied the interaction of lithium and magnesium with adenine nucleotides (see Sec. 4.2.1.). These chemical models of lithium's action presuppose that there is a direct action of the metal ion on a specific site or group of sites in some receptor-effector molecule. Frausto da Silva and Williams [60] have argued that one should regard the biological system as a whole, rather than each individual complex in isolation, and taking this view one could envisage that the various chelating ligands present--for instance, nucleotides--might sequester otherwise active ions such as magnesium to leave available sites for occupation by the lithium, which has a much lower affinity for such sites. This scenario assumes that there are more cation binding sites in the biological system than cations competing for them and that a small change in the concentration of high-affinity sites might cause a disproportionate change in the population of different metals which inhabit the lower-affinity sites. Frausto da Silva and Williams therefore suggest the use of "conditional" association constants in the determination of such interactions rather than the classical constants determined in simple aqueous solutions.

Of course, it is possible that the above argument might be inverted. Were lithium to bind at some particularly high-affinity site which would have otherwise been occupied by another metal--for instance, magnesium--this would affect the distribution of magnesium in other systems. In this way it could be envisaged that effects of lithium seen are, in fact, a consequence of the redistribution of magnesium rather than a direct effect on some system of lithium itself.

The initiation of an event by the conformational change induced by metal binding ultimately must be reversed by the release of the metal from its binding site. Such variations in binding and release might be induced by the local molecular environment: local changes in the concentration of other ions, including protons, may be the regulatory factors. Since weakly binding species are more susceptible to small changes in local conditions, it may be that fine regulation is achieved by small, and therefore metabolically inexpensive, changes in metal concentrations. If this were true, one could readily see that another weakly binding species, lithium, could perturb the system sufficiently to have widespread effects and that it might affect the binding-release characteristics of biologically active ligands with other metals.

It is to be hoped that the increased interest in the weak interactions of alkali metals with various ligands will provide the background for a more rational understanding of the molecular pharmacology of lithium.

4. PHARMACOLOGICAL AND BIOCHEMICAL ASPECTS OF LITHIUM

The pharmacological and biochemical studies on lithium fall into a number of categories, many of which have no relation to its present use in psychiatry. It is essential, therefore, to sound a note of caution against extrapolation from chemical and experimental conditions which are not those obtained during lithium therapy. Very early biochemical literature is based on analysis of lithium distribution, and this was further extended immediately following the reports of intoxication from the United States [61]. These studies led shortly to a number of purely toxicological investigations when high doses of lithium were given to animals.

In parallel with these early studies are those in which lithium was used as substitute for sodium. When a substance is used to replace a naturally occurring constituent of a physiological solution the new substance must be shown to have no pharmacological action on the system. For lithium the original assumption was untested, and the logic of many papers appears to be: "Lithium = sodium; therefore, we can replace some or all of the sodium with lithium; however, lithium is *not* sodium and therefore any effects which we see are due to the lack of sodium." Based on this argument a large number of studies have been carried out in which lithium isotonically replaced

sodium in physiological solutions (such as Ringer's solution) to a
concentration of 150 mmol/liter. Tissues and organs were then perfused in these solutions and the results interpreted as described
above. I would remind the reader that the approximate concentration
of lithium in the tissues and fluids of treated manic-depressive
patients is about 1 mmol/liter. Similarly, lithium has been used
for the study of cation effects on enzymes at 100 mmol/liter and in
disaggregation of ribosomes at 1,000 mmol/liter. Finally there are
studies which have as their aim the investigation of the psychiatric
effect of lithium but which have not recognized that biology does
not take place in a *simple* aqueous environment. The biological
system is multicompartmental, containing a large range of macromolecules, and the interaction between these ligands and between the
ligands and the aqueous phase is extremely sensitive to the ionic
constituents.

In considering the effects of lithium on whole animals we must
also be aware that lithium is not immediately lethal except at very
high concentrations and that its toxicity derives primarily from es-
calating failure of the kidney to excrete the ion. Short-term
studies in animals, therefore, using doses of lithium which would in
the long term be fatal cannot be extrapolated with confidence to explain pharmacological events during long-term lithium therapy.

All these different types of study are retrieved in literature
searches and may be adduced uncritically in support of mechanistic
theories.

The pharmacological and biochemical literature until 1977 has
been reviewed by the present author [62,63] and by others [64-66]
and may be found in the compendia listed in Table 1. The function
of the present review is to update these earlier surveys and to discuss in detail some aspects of lithium which have become of prime
interest.

4.1. Absorption and Distribution

Lithium is usually administered to patients in the form of lithium
carbonate tablets of between 250 and 400 mg to give a total of between 250 and 2,000 mg/day. In animal studies it is easier to give
the daily dose either in the drinking fluid or in the diet, though
some studies have been carried out with the dose given by intraperitoneal injection. The latter produces a much sharper rise in plasma
lithium and consequent rapid excretion so that the exposure throughout the day is much lower than that of either of the oral routes
[67]. Interspecific variation occurs, and it should not be inferred
that a similar dosage with respect to body weight given to a rat,
for instance, will produce a mean plasma lithium in the same range
as that of the human since the rat kidney conserves water better
than the human and the dose required to keep plasma concentration of
lithium high is so large that renal accumulation occurs. Such
methodological difficulties have been discussed by Thomsen et al.
[68].

        Lithium is only given to humans by the oral route, by far the
simplest, and there is no justification for administration by other,
more hazardous routes.

4.1.1. "Slow release" versus Conventional Preparations

The aim of lithium therapy has been to maintain a steady plasma
lithium throughout the day since it is held that many of the side
effects of lithium result from relatively sharp peaks in plasma
lithium concentration following absorption of a single dose. Since
lithium carbonate itself is not patentable, some drug companies have
been eager to introduce slow-release formulations which will obviate
the plasma peaks. A wide range of such formulations are now available [69]. Of the 52 preparations listed by Schou, one-quarter are
claimed to have "slow-release" action. Very few of these preparations are available except in the country of origin and its near
neighbors.

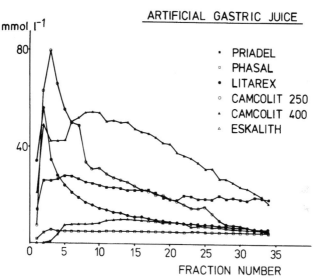

FIG. 5. Dissolution of various commercial preparations of lithium in simulated gastric juice (0.1 mol/liter HCl). Priadel, Phasal, and Litarex are "slow-release" formulations, while the remainder are preparations of lithium carbonate B.P.* (From N. J. Birch, *Inorg. Persp. Biol. Med.*, *1*, 173, 1978, reproduced by permission of the copyright holder.)

Considerable controversy surrounds the question whether or not slow-release lithium preparations are of any significant value in lithium prophylaxis. The advantages claimed for such slow-release preparations are that they reduce the incidence of side effects because they reduce the height of the peak in plasma lithium following a single dose. Until recently standards for slow-release preparations depended upon dissolution tests such as that defined in the *United States Pharmacopoeia* (19th edition). In this test tablets are placed in a small basket and agitated in an artificial gastric or intestinal fluid to simulate conditions which might be expected as the tablet passes down the gastrointestinal tract. Aliquots of the

*B.P. is the abbreviation for "British Pharmacopoeia" and indicates that the drug has been formulated according to criteria laid down in the B.P.

This applies to all other occurrences of "B.P."

(a)

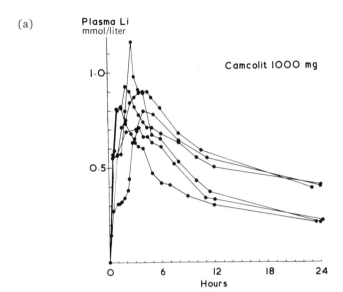

(b)

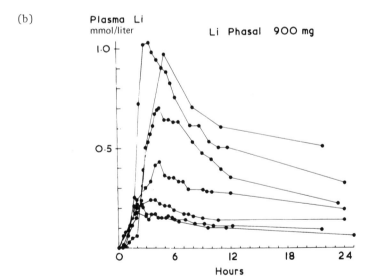

FIG. 6. Plasma concentrations of lithium after administration of different preparations of lithium to normal human volunteers. Formulation (a), Camcolit, is conventional lithium carbonate B.P., while the two other preparations (b), (c), are marketed as "slow-release." (From N. J. Birch, *Inorg. Pers. Biol. Med.*, *1*, 173, 1978, reproduced by permission of the copyright holder.)

(c)

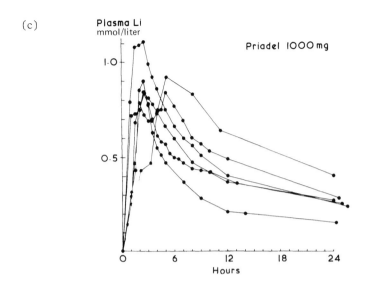

FIG. 6 (continued)

external fluid are taken at intervals and analyzed for lithium [70].
We have carried out a different test in which a tablet is placed in
a small chamber with sintered glass base through which is pumped the
artifical solution. The eluate from this perfusion chamber is collected in a fraction collector in small aliquots and these are
analyzed. This latter test gives an indication of the rate of release of lithium at more discrete intervals than does the official
test. It is clear from this that the supposed slow-release preparations do indeed dissolve slowly (Fig. 5). However, a better measure
of the bioavailability of the preparations is given by following the
profile of serum lithium in a human subject following a single dose,
and we have carried this out for a number of preparations (Fig. 6).
It is clear from this that the slow-release preparations are considerably more variable between subjects than the conventional
lithium carbonate. This was confirmed when we considered the amount
of lithium recovered from the feces of such subjects [70]. The
preparations which dissolve slowly in vitro also dissolve slowly in
vivo, but at a rate which is variable between subjects. We feel that
this variability is undesirable since changes in gastrointestinal

function may lead to changes in lithium absorbed, with consequent increase or decrease of plasma lithium [70,71].

It is the view of our particular research group that it is better to provide lithium in smaller doses two or three times each day than to give a single dose of a slow-release preparation, since not only will the smaller dose be more readily tolerated and produce fewer side effects as a result of the smaller plasma peak, but in the long term a more steady plasma lithium concentration can be maintained. These matters have been discussed in great detail by Johnson [72], who has reviewed widely the varying points of view. He concludes that "there is no direct answer to the question of which type of lithium product is to be preferred, . . ." but goes on to suggest that slow-release formulations are best for patients who are prone to side effects. Naturally we disagree with this assessment since if there is an area of doubt, the simpler preparation is to be preferred. On a more general level it may be argued that genuine slow-release preparations can only be produced for drugs which have either a saturable rate of absorption, that is, some carrier mechanism, or which can be presented in an inactive form to a saturable mechanism regulating the gastrointestinal absorption of the inactive drug or its metabolism to the active form within the body. All other preparations, as in the case of lithium, depend on the particular conditions which occur in the gut following administration and will therefore vary between subjects and at different times in the same subject.

4.1.2. Distribution in Tissues

During prophylactic treatment with lithium the concentration in most tissues is similar to that of plasma, though the element is accumulated in bone [73] and thyroid [74]. The accumulation of lithium in bone was originally shown by Desgrez and Meunier [75]. Recent work has suggested that lithium is incorporated into the mineral, where it is relatively inaccessible, though about 40% of bone lithium occurs in the bone water whence it is readily removed [73]. Lithium accumulation in thyroid gland is of particular interest since one of the known side effects of lithium therapy is hypothyroidism.

TABLE 4

Concentration of Lithium in Different Tissues
with a Variety of Treatment Regimes

Ref:	76	77	77	78	79	80	80
Type of Expt.:	Acute	Acute	Chronic (12-day)	Chronic (18-month)	Chronic (28-day)	Li depletion	Control
Route of admin.:	i.v.	i.p.	i.p.	d.f.	d.f.	Li-free diet	Control diet
Daily Li intake (mmol/kg):	7.2	0.6	0.6	0.8	0.8	0	Unknown
Time after last dose (hr):	24	12	12	12	12	2nd Generation	
Tissue	mmol/kg wet wt					μmol/kg wet wt	
Brain	1.22			0.18	0.18	1.86	8.14
Muscle	1.60			0.30	0.22	<4.3	30.00
Bone	1.59			1.38	1.15	0.10	2.57
Liver	0.53			0.15		0.63	15.71
Kidney	1.35			0.53		1.19	6.57
Heart	0.97					1.40	6.57
Spleen	0.82					1.40	4.14
Serum/ plasma	1.22	0.08	0.21	0.17	0.14	0.42	1.17
Pituitary		0.26	0.39			18.6	21.43
Thyroid		0.31	0.43				
Adrenal		0.15	0.24			5.0	5.43

Key: i.v. = intravenous; i.p. = intraperitoneal; d.f. = drinking fluid.

Table 4 shows the concentrations of lithium reported from various tissues. Though these concentrations are not markedly different from plasma concentrations, they can be seen to be related to the rate at which lithium enters and leaves particular tissues. Schou reported an extensive series of studies on the tissue kinetics and found that lithium passed rapidly from blood into the kidney, more slowly into liver, bone, and muscle, and very slowly indeed into brain. Liver and muscle reached approximate equilibrium within the 48 hr studied, while brain, kidney, and bone had not established a steady state by the end of this period. This goes some way to explain the slow onset of the prophylactic effect of lithium.

We have recently carried out studies using the stable isotope 6-lithium in rats which were previously treated with the predominant naturally occurring isotope 7-lithium. These studies have broadly confirmed Schou's finding and have shown that the presence of lithium in the tissue does not affect the rate of uptake of the element. 6-Lithium was determined by atomic absorption spectrometry [81], but we have been unable to determine the element in the presence of high concentrations of calcium and results from bone are therefore not available.

In an attempt to identify a locus of action in the brain the distribution in various brain areas has been determined by a number of workers [76,82-84]. A particularly interesting technique has recently been reported for the direct determination of lithium distribution in tissues by the neutron activation of a histological section of animals previously treated with 6-lithium [85-87]. This latter technique has enabled the direct identification of localization of lithium without the intrinsic errors of microdissection of anatomical regions.

Broadly, the classical dissection and dissolution techniques agree with the results found by Thellier and his colleagues, though it must be stressed that each of these studies used a different treatment regime and for the reasons previously discussed the tissue concentrations cannot be claimed to be representative of those occurring in the human brain after long-term lithium. There is general

agreement that lithium is accumulated particularly in the basal gan-
glia, the hippocampus, the hypothalamus, and the cerebral cortex.
Thellier and his colleagues also report an increased concentration
in the cerebellum. The accumulation in the hippocampus is of interest because of the suggestion that the limbic system may be involved in the regulation of emotion. The accumulation in basal gan-
glia may be relevant to the suggested use of lithium in the treatment
of movement disorders. The micrographs published by Thellier and
colleagues give an interesting overall view of the lithium distribution, and it is clear that some areas are not readily accessible to
lithium, particularly the corpus callosum and the thalamus. However,
direct visualization of the distribution brings to mind the thought
that lithium content may vary in terms of the distribution of water
within the brain and that lithium has difficulty in penetrating the
more lipid regions.

Ebadi et al. [76] also showed an accumulation of lithium in
the pituitary gland, and this is of particular importance when one
recognizes that this endocrine organ is the major interface between
the brain, via the hypothalamus, and the rest of the endocrine regulatory system. Stern et al. [77] have confirmed that the pituitary
accumulates lithium as do both the thyroid and adrenal glands.
Since these three glands are the major endocrine regulators of the
body, this is of particular interest in the search for the mode of
action of lithium.

A provocative recent finding has been that of Patt et al. [80]
who have investigated the possibility that lithium may be an essential element in the rat and have maintained animals for three gener-
ations on lithium-depleted diet. After one generation the pituitary
gland contained 10 times more lithium than any other tissue with the
exception of the adrenal gland which had approximately one-half the
content of the pituitary. Both the pituitary and the adrenals contained similar quantity of lithium in the depleted group and the
group treated with commercial diets, presumably of average lithium
content. In the control group, bone had the highest content of
lithium and might be considered to be a measure of lithium exposure.

This latter is in general agreement with other studies on bone [73].
In second-generation lithium-depleted animals the pituitary had the
same concentration of lithium as both the first-generation lithium-
depleted and both sets of controls. The second-generation animals
had somewhat reduced adrenal gland concentration, but it was still
higher than the remainder of the tissues. These findings suggest
that the pituitary and adrenal glands accumulate lithium and may indeed protect themselves from lithium depletion. The second- and
third-generation females had reduced fertility compared with the
first generation and the normal control group, although there was
no effect on the growth rate. It should be noted in passing that
lithium treatment of rats resulted in decreased brain concentration
of magnesium [79,82].

Whether these animal studies, almost exclusively short-term,
reflect the distribution of lithium in the manic-depressive patient
is not completely known, and of course this information will not
readily be obtained. There is the possibility that long-term treatment with lithium may affect the way in which the body absorbs and
distributes the ion, and for this reason our recently developed technique using the stable isotope 6-lithium has particular potential
[81]. Since the isotope is not radioactive (indeed, there are no
useful radioactive lithium isotopes) it is possible for it to be administered to human subjects without the ethical problems of radio-
isotope administration. We have attempted in normal volunteers who
were preloaded with lithium for several days to follow the kinetics
of a single dose of 6-lithium, and the results suggest that the
plasma profile of 6-lithium in the presence of a preexisting lithium
load is identical to the profile seen in the untreated subject.
Figure 7 shows that the plasma concentration of 6-lithium is super-
imposed upon the declining 7-lithium concentration curve from the
previous loading dose. The shape of the 6-lithium curve can be seen
to be similar to that in Fig. 6(a), which was the preparation used
to preload the subjects.

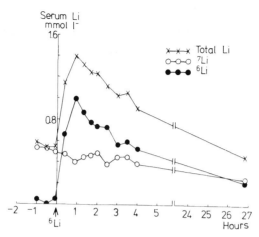

FIG. 7. Plasma concentrations of ^6Li and ^7Li following a single dose of $^6Li_2CO_3$ (1 g). The normal human volunteer had previously received 250 mg Camcolit, eight hourly, for 4 days. (From Ref. 81, with permission of the copyright holder.)

4.2. Subcellular Actions of Lithium

Much of the biochemical work on lithium has been carried out with no direct interest in the mode of action in psychiatric disorders. Lithium frequently has been used as one of a series of interchangeable metal ions having no regard to their intrinsic properties, but merely as suitable packages of charge of known size. Alternatively, lithium has been used as a reagent to perform separations [88].

   In mitochondria lithium was compared with sodium, potassium, magnesium, and Tris in the study of ATP and ADP exchanges when no difference was seen between the effects of the different cations [89]. At low concentrations of lithium in rat brain mitochondria ADP-deficient respiration was inhibited at below 10 mmol/liter lithium but stimulated above this concentration [90]. Other workers report that in rat heart mitochondria lithium caused a stimulation of calcium efflux, though this was not as effective as the stimulation caused by sodium [91].

   At below 1 mmol lithium promotes tubulin polymerization in the presence of magnesium, and this may have a role in the stabilization of microtubules in neurosecretory systems [92].

A large literature exists on the relationship between lithium
and the synthesis and secretion of biogenic amines and the response
of neural tissues to these neurotransmitters. Shaw [93] has reviewed this field but concludes that the variety of experimental
conditions and pharmacological insults make it very difficult to say
which are the most significant, or even valid, observations and that
many of the results may be secondary to toxicity. Other reviewers
take a more optimistic view that such studies are indeed indicative
of the mode of action [94-98]. The variety of data so far produced
has, however, not allowed any clear conclusion and it could be argued
that the systems studied have historically been those which had ready
analytical techniques for the transmitters studied. As further analytical techniques have been developed the range of lithium effects has
been widened. Since there are now over 30 different neurotransmitter
systems known in the brain, it seems unlikely that by chance one has
stumbled across the major site of action in the few systems which
have been studied so far. The contradictory nature of the literature
lends weight to this interpretation. However, the amine hypotheses
maintain much currency and in some quarters their exclusive study
has resulted in the neglect of alternative approaches. However, the
present reviewer can do no more than direct the reader toward the
very detailed reviews cited and say that, whatever the mode of action
of lithium may prove to be, it must also explain these diverse
findings.

It could be argued that the simplicity of the lithium ion and
the diversity of its effects suggest that the site of action is at a
very fundamental level, and with this in mind we will now consider
two such systems.

4.2.1. Effects on Enzymes and Cofactors

Table 5 lists a range of enzymes on which lithium has been shown to
have an effect. The diversity of these systems underlines the apparently nonspecific nature of lithium action, and indeed the enzymes
listed range from those which are purely autonomous and regulated by

TABLE 5

Enzymes Affected by Lithium (with
Representative References)

Glycolysis	*Nucleic acid synthesis*
Hexokinase [99]	RNA synthetase [106]
Glucokinase [100]	DNA polymerase [107
Phosphofructokinase [101]	
Fructose 1-6-diphosphatase [102,103]	*Miscellaneous*
Enolase [103,104]	Cholinesterase [111]
Pyruvate kinase [105]	Acetyl cholinesterase [112]
	Aryl sulfatase [111]
Krebs cycle	Alkaline phosphatase [111]
	Acid phosphatase [111]
Aconitase [109]	Na^+/K^+ ATPase [113]
Succinate dehydrogenase [110]	Mg^{2+} ATPase [114]
	Ca^{2+} ATPase [115]
Amino acid metabolism	
Tyrosine aminotransferase [100]	
Alanine aminotransferase [108]	
Tryptophan oxygenase [100]	

substrate or product through those which are induced and finally include those which control the expression and replication of the
genetic regulators of the cell and hence control the activity of
wider systems.

We have chosen to study glycolysis because of its very fundamental position in metabolism and because many of its enzymes are
magnesium-dependent. This does not imply that glycolysis is the
major locus of action of lithium, but the system does provide a well-
characterized model in which many of the regulatory steps are known.

Glycolysis has particular importance in brain metabolism since
under normal circumstances glucose is the major energy source of the
brain. We have investigated a number of the independent enzymes in

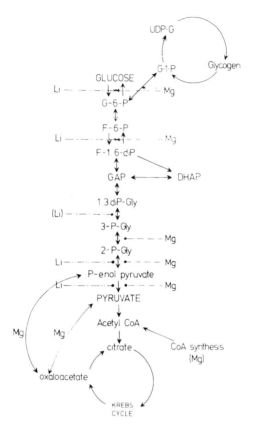

FIG. 8. Glycolysis, showing magnesium-dependent steps and enzymes on which lithium action has been investigated.

the cycle (see Fig. 8). In particular we have shown that lithium inhibits pyruvate kinase at pharmacological concentrations both in enzyme obtained from rabbit muscle and from a purified rat brain preparation [105]. Inhibition characteristics of these two different
sources were identical, and lithium was found to be competitive with
respect to ADP and noncompetitive with respect to the other substrates: phosphoenol pyruvate, potassium, and magnesium. Pyruvate
kinase is also inhibited by calcium, and its characteristics are
identical except that calcium inhibition is competitive with respect
to magnesium. At this stage we considered the possibility that

lithium might interfere in the magnesium-ADP complex which is substrate for the reaction, and we determined by chromatographic techniques the stability constants for ATP and ADP complexes with both
lithium and magnesium [116]. It seemed likely from the data obtained
that lithium, although able to form a complex with both nucleotides,
was unable to displace magnesium from such sites.

Recent studies by Mildvan and his colleagues (summarized in
Ref. 117) have suggested that pyruvate kinase requires a monovalent
cation, usually potassium, and that there are two distinct divalent
cation sites which may be occupied by the same or different metals.
In vivo these are assumed both to be magnesium. Since it is possible
to replace magnesium in these sites by manganese, and the latter
would be a suitable paramagnetic probe, we set out to identify the
site of action of lithium in pyruvate kinase. We had already shown
that lithium was noncompetitive with regard to potassium, and this
suggested that one or both of the divalent sites was involved. We
had intended to carry out electron paramagnetic resonance (EPR)
studies of manganese-activated enzyme and the effects thereon of
lithium, but the preliminary studies of the effects of lithium on
manganese-activated pyruvate kinase were unexpected and provided a
new perspective [118]. Lithium did not inhibit manganese-activated
pyruvate kinase while calcium did (Table 6). This finding suggests
that lithium and calcium inhibit at different sites.

We have carried out further studies of the binding of lithium
and magnesium to ATP and ADP using nuclear magnetic resonance (NMR)
[119] and have confirmed our original interpretation that lithium is
able to bind to the nucleotide but is readily displaced by magnesium.
We have further shown that there appear to be two sites for magnesium
binding to the nucleotide, one of which is saturated at about 1:1
stoichiometric ratio and the other which appears to progressively
fill once the first site is saturated. Lithium binding appears to
be distinct from either of these sites.

Presently we interpret all these data in the following manner.
The inhibition of lithium appears to take place on the divalent
metal site at the enzyme surface to which attaches, during the

TABLE 6

Inhibition Characteristics of Rabbit
Muscle Pyruvate Kinase

Activator:	Mg		Mn	
Inhibitor:	Li	Ca	Li	Ca
Substrate				
Phosphoenol Pyruvate	n.c.	n.c.	(-)	n.c.
K^+	n.c.	n.c.	(-)	n.c.
Mg^{2+}	n.c.	comp.	(-)	comp.
ADP	comp.	comp.	(-)	comp.

Key: Lithium or calcium inhibition was competitive (comp.) or non-competitive (n.c.) with respect to the various substrates shown; (-) indicates no inhibition.

Source: N. J. Birch and P. K. Kajda, unpublished results.

phosphoryl transfer, the divalent metal-nucleotide ($ADP-M^{2+}$) complex. Our NMR data suggest that lithium is not involved in the $ADP-M^{2+}$ complex and our kinetic data suggest that the monovalent site is not available. Since at physiological pH the majority of ADP occurs as magnesium-ADP complex, this would be in agreement with the noncompetitive nature of lithium inhibition with respect to magnesium and its competition with ADP. The presence of manganese prevents the conformational change associated with lithium. On the other hand, calcium may act at either divalent site, and this is consistent with its kinetic behavior in both magnesium- and manganese-activated enzyme. Further studies are now in progress to determine whether lithium displaces magnesium or whether it causes a conformational change which prevents the binding of magnesium.

We have carried out a number of studies of other magnesium-dependent glycolytic enzymes. Hexokinase was not inhibited at pharmacological concentrations of lithium. Phosphofructokinase (PFK), however, was inhibited, and this inhibition was competitive with respect to ADP and magnesium but noncompetitive with respect to

potassium and fructose-6-phosphate [101]. Since PFK is the major
regulatory step in glycolysis, this finding is significant, though
the degree of inhibition was rather small. In contrast, the inhibition by lithium of the reverse reaction to PFK, that of fructose-1-
6-diphosphatase, was 47% at 2 mmol/liter lithium, and this may suggest that lithium disturbs the relationship between glycolysis and
gluconeogenesis or perhaps prevents the operation of futile cycles
at this point. We have also shown inhibition by lithium of enolase
and phosphoglycerate kinase, though these and the remaining enzymes
have yet to be studied in detail.

Having shown such inhibition of individual enzymes it was important to establish that lithium did indeed have an effect on glycolysis as a whole. We have therefore undertaken preliminary studies
on glucose utilization in human erythrocytes [120]. Lithium does
appear to inhibit the utilization of glucose, and these studies are
being continued.

The studies on lithium and glycolysis are put into perspective
by previous reports of changes in carbohydrate metabolism in patients
receiving lithium [121-123] and studies on glucose utilization and
glycogen synthesis in rats [124]. These studies suggest that there
is an increase in glucose utilization and glycogen synthesis, though
there is disagreement with regard to the mechanisms involved. The
importance of ions in the regulation of metabolism has been extensively reviewed by Rasmussen and his colleagues [125].

4.2.2. Cyclic-AMP System

The action of many hormones at the cellular level involves cyclic
adenosine 3',5'-monophosphate as the intracellular mediator or
"second messenger." The cyclic nucleotide is synthesized from ATP
by adenyl cyclase, which is magnesium-dependent. Differences occur
in specificity between tissues in their relevant hormonal or neurotransmitter activators, but this specificity does not depend on the
catalytic subunit, which appears to be identical in the variety of
tissues, but on a specific regulatory subunit activated only by

particular neurohumors [126]. In the central nervous system a number of supposed neurotransmitters have been identified as activators of adenyl cyclase, including noradrenaline, dopamine, and histamine. Cyclic-AMP is also regulated by the rate of degradation by the specific phosphodiesterase, which is also magnesium-dependent.

Because of its pivotal role in neuroendocrine regulation, adenyl cyclase and cyclic-AMP have provided a focus for those studying the affective disorders and lithium. A number of miscellaneous effects of lithium have been reported, and adenyl cyclase-cyclic-AMP interactions provide an attractive unifying hypothesis. Three recent reviews [126-128] have evaluated the literature, and I will therefore provide only signposts to current research activity.

Adenyl cyclase is affected by a number of hormones, and the effects of these may be investigated both in humans and in animals, though there has been some conflict in the data from these two sources. Geisler et al. have investigated a number of aspects of cyclic-AMP and adenyl cyclase and found that cyclic-AMP in the cerebrospinal fluid (CSF) of manic-depressive patients was not different from normals nor from other psychiatric patients [129], nor was there any correlation between the mood rating and the CSF cyclic-AMP, the use of electroconvulsive therapy, or lithium. Ebstein and Belmaker [126] were unable to show any effect of lithium on glucagon-induced rise in plasma cyclic-AMP in humans but showed inhibition of the cyclic-AMP response to adrenaline.

Animal studies have shown other results; for instance, Christensen and Geisler [130] demonstrated effects on vasopressin-sensitive cyclic-AMP and its excretion in rats. In tissue studies Schimmer [131,132] reported enhancement by lithium of catecholamine-stimulated adenyl cyclase in cultured glial tumor cells. Murphy et al. [133] and Wang et al. [134] investigated lithium actions on prostaglandin-stimulated adenyl cyclase, while Spiegel et al. [135] and Steele [136] have shown inhibition of parathormone-stimulated adenyl cyclase. Zatz [137] has shown inhibition of release of cyclic-AMP in the pineal gland which was inversely proportional to the magnesium concentration.

The results thus divide into two main groups: Lithium has an effect, mainly inhibitory, on the cellular function of adenyl cyclase under a variety of hormonal and neurohumoral stimuli; the clinical results are much less convincing, and it is likely that monitoring the metabolism of cyclic-AMP by measurement of its plasma and urinary concentrations is too remote a technique and insensitive to cellular events. It may be that the effect of lithium on second messenger system proves to be the unifying mechanism by which its widespread effects are produced at the varying levels of metabolic organization of the body. However, it is my personal view that the cyclic-AMP-related phenomena are a reflection of a fundamental effect of lithium on magnesium-dependent enzymes.

4.3. Membrane Effects

It has been suggested that the concentration of lithium in the red blood cells (erythrocytes) during lithium prophylactic treatment may have clinical value either in the prediction of brain lithium concentrations [138] or for discrimination between unipolar and bipolar patients and responders and nonresponders to the drug [139,140]. These proposals have been widely disputed [141,142].

The kinetics of lithium uptake by normal red cells has been studied in detail. The most definitive work to identify the transport mechanisms has been carried out by two groups, in Munich, West Germany and in Chicago and Harvard in the United States [143-146]. They have identified four different components of lithium transport in the red cell:

1. Ouabain-sensitive uptake of lithium by the sodium-potassium ATPase system
2. "Leak" diffusion of lithium
3. Phloretin-sensitive uptake resulting from an electro-genic sodium-lithium countertransport
4. An anion exchange where lithium is transported as a result of ion pairing with bicarbonate

Aspects of these transport systems have been confirmed by other workers [147,148]. Meltzer et al. [149] report that though the lithium pump activity in manic-depressives is not different from normals prior to lithium treatment, its activity decreases after several days of lithium administration, which suggests that the ion has a long-term effect on the membrane structure and function. This is supported by a study indicating an irreversible inhibition of choline uptake by erythrocytes following lithium [150,151]. Dorus et al. [146] have identified a genetically related effect on lithium transport mechanism which is present in the families of some manic-depressive psychotics. Canessa et al. [152] have shown an increased sodium-lithium countertransport to be typical of essential hypertension: if this is confirmed, a major advance has been made in the detection of this disease.

Two recent studies have shown decreased uptake of 5-hydroxytryptamine by human blood platelets which defect was reversed by long-term treatment by lithium [153,154].

Lithium stimulates sodium-potassium ATPase in the human red cell [155,156] to cause sodium efflux which is ouabain-sensitive and stimulated by sodium, potassium, and lithium. Glen [157] confirms that lithium has a potassiumlike action on the potassium-sensitive side of the membrane and concludes that there is substantial evidence to support the view that lithium acts to stimulate transport during depression and inhibit it during mania. A recent report of magnesium-activated and calcium/magnesium-activated ATPase stimulation by lithium in rat iris and visual cortex [158] suggests that lithium's action is not limited to the sodium pump but may be seen in other ATPases, as had previously been reported [114,115,159].

4.4. Effects of Lithium on Tissues, Organs, and Systems

It is clear from the foregoing sections that the effects of lithium are legion, and in this section I intend merely to point to sources of information and to highlight the areas of current controversy.

4.4.1. *Thyroid Gland*

Hypothyroidism is a frequently reported side effect during lithium
treatment and has been reviewed by Hullin [160,161] who has reported
that female patients, especially those over 40 years of age, are
more susceptible to lithium-induced thyroid effects than males,
though this may not necessarily be a result of lithium therapy since
women of this age group also have a higher incidence of hypothyroid
symptoms generally. Lithium does, however, undoubtedly have effects
on thyroid function, and Wolff [162] has reported lithium inhibition
of thyroid adenyl cyclase which was particularly sensitive to the
magnesium concentration of the medium. Lithium also interferes with
the coupling of iodotyrosines to form the iodothyronine hormones,
triodothyronine (T_3), and thyroxine (T_4) [163,164] and also with the
release of the hormone from the gland [92]. The stimulating effect
of thyroid-stimulating hormone (TSH) on the gland is also inhibited
[165], and this may result in the compensatory increase in TSH often
seen following lithium [160].

The accumulation of lithium in the thyroid gland may result in
the amplification of lithium effects on transport mechanisms in the
gland with consequent effects on the secretion of the thyroid
hormones. Lithium inhibits iodide transport within the thyroid
gland, and this enhances the goitrous response. Patients showing
hypothyroidism following lithium are readily treated with small doses
of thyroxine.

4.4.2. *Bone*

Lithium is accumulated in bone to a higher concentration than in most
other tissues [78], and there was some concern that such accumulation
might have effects on the structure and function of bone in long-
term-treated patients [79]. Indeed, Christiansen and his colleagues
[166] have reported decreased bone mineral content in patients receiving lithium. However, later studies suggested that these differences were a result of the different disease states studied and
that bipolar manic-depressive patients showed reduced bone mineral

content while unipolar patients showed no change [167]. We have
carried out an extensive series of studies in lithium-treated
patients and in rats and we conclude that any effects which we have
seen have been the result of the normal aging process and not an
effect of lithium [160,168]. In rats a decrease in bone size was
associated with decreased body weight and for this reason caution
should be used in the treatment of persons of immature bone
structure.

4.4.3. Kidney

One of the most contentious issues in the field of lithium research
since the mid-1970s has been the possibility of long-term damage to
the kidney following lithium treatment. It was reported that a
number of long-term-treated lithium patients had developed serious
damage which was detected on biopsy of the kidney when apparently
characteristic biological lesions were seen in a high proportion of
patients [169-171]. Lithium had been reported for many years to be
associated with polyuria, and in some areas this had been considered
to be an almost inevitable side effect which was well tolerated by
the patients. Lithium previously had been shown to cause inhibition
of antidiuretic hormone action [172] and it was assumed that the
polyuria was a consequence of this.

The appearance of the histological lesions in patients who
were demonstrating polyuria gave rise to the misconception that
these two phenomena were linked. However, polyuria is completely
nonspecific and it may be seen in a large number of patients for
different reasons. In subnormal children, for instance, it is a
matter of nursing lore that the children drink a lot of fluid and
this is reflected in their urine output. Similarly, for behavioral
reasons other patients may indulge in excessive drinking. Indeed,
we discovered a patient in a neighboring ward to our research unit
who produced 25 liters of urine per day, who had not been detected
because his bladder capacity had developed over the years so that he
was able to void 5 liters of urine on each micturition. This was

entirely polydipsia-induced polyuria. The patient had never received lithium and his renal function was otherwise normal.

In the early days of lithium therapy, because of the dangers
appreciated of toxicity, it was the common practice to insist that
patients ensure an adequate water intake, and I believe that this
instruction, together with the lack of counterinstruction, has led
to the resetting of the thirst mechanism by the persistent habit of
drinking. That such factors as habit are important in the regulation of fluid intake was recognized in a review of the control of
thirst mechanisms [173]. A further factor in inducing increased
fluid intake might be that some patients find the taste of lithium
to be unpleasant and therefore drink to remove the taste. Some
measure of the variety of such "lithium-induced polyuria" may be
seen from the report that 40% of all patients in some centers show a
daily urine volume greater than 3 liters while in others the incidence is about 4% [174].

As a result of the reported nephrotoxic effect of lithium the
renal physiology of lithium has been well documented [172,175] and a
special symposium was organized [176].

Two factors are not in doubt: The histological lesions do
exist in some patients following lithium therapy, and it is clear
lithium does have effects on renal physiology [177]. However, the
location of reports of nephrotoxic effects also coincides with a
high incidence of lithium in toxicity reported. A review on lithium
intoxication [178] was able to report the treatment of 23 cases in
one center; indeed, most of the cases of the histological lesions
first reported had at some stage suffered lithium intoxication [169].
It is the experience of our lithium clinic, some 150 patients, over
the period 1970-1980 that only two cases of intoxication have been
seen and these were suicide attempts. We believe that *lithium intoxication can be prevented by* using relatively *low lithium doses
and by regular monitoring*. We have studied in detail the renal
function of our patients and conclude that no decrease is observed,
although the secretion of the antidiuretic hormone following lithium

is markedly elevated [179]. It is our contention that the histological changes demonstrated are indicative of previous lithium
intoxication. It can be demonstrated that the renal concentrating
ability, although reported to be decreased by lithium [177], is not
impaired permanently, and this is tested ultimately by the report by
Glen et al. [180] that in 784 patients receiving lithium over a 10-
year period in Edinburgh 33 died during the period of study, of
which none could be attributed to renal damage. The one case reported of uremia had suffered chronic pyelonephritis for several
years prior to starting lithium. The deaths were due predominantly
to suicide or cardiovascular disease, and bearing in mind that the
population studied was relatively elderly, the death rate is almost
exactly that of a comparable population of untreated manic-depressives.

In studies of rats Christensen and Hansen [181] have demonstrated that impaired renal concentrating ability following lithium
was reversed on discontinuation of the drug.

The mechanism of lithium excretion has been discussed in the
previous section on intoxication and has been reviewed in depth [172,
182]. The endocrine consequences of the lithium-induced sodium loss
have been summarized [183] and parallel changes in hydrion excretion
in the tubule have been reported [184]. The effect of lithium on
the sodium reabsorption might be due to direct competition for transport or to an interaction with receptors for antidiuretic hormone or
aldosterone. In vitro, lithium inhibits aldosterone-induced sodium
transport, but it is possible that in the long-term-treated patient
the interaction may be more complex and involve the renin-angiotensin-
aldosterone regulatory system. Changes in aldosterone production have
been reported at the "switch" period in short-cycle manic-depressive
patients [34]. In isolated perfused rat kidneys lithium inhibited
the reabsorption of both sodium and calcium, though there was no effect on the transport of magnesium, potassium, or inorganic phosphate
[185]. One curious feature of these in vitro studies was that lithium
was retained by the kidney even after perfusion with lithium-free
solutions. This retention of lithium might account for the persistence of lithium effects in the kidney.

4.4.4. Nervous System

Experiments on the intact nervous system are difficult and there is
the problem that, while experiments cannot be carried out directly
on patients suffering from affective disorders, there is no animal
model which will allow predictive experiments. Biochemical studies
have been performed on cells and fractions from brains of lithium-
treated animals, but there is still the problem of interpretation
and extrapolation to the clinical situation. The direct study of
the electroencephalogram (EEG) has been so far unrewarding [186].
A 48-hr periodic psychotic patient, discussed in an earlier section,
was extensively studied with EEG, and although lithium resulted in a
generalized slowing of the EEG pattern with increased amplitude,
this was so nonspecific that no conclusion can be drawn of mode of
action [186].

The neural and neuromuscular side effects of lithium have received some attention [187], primarily with a view to detecting neuro-
toxicity of lithium at an early stage. It is claimed that lithium increases the depth and length of sleep, and this can be detected on the
sleeping EEG [187]. Although lithium has been reported to cause muscular weakness, few studies have been carried out on this aspect although electromyographic (EMG) studies have suggested that lithium
reduced maximal motor nerve conduction velocity [188].

4.4.5. Cardiovascular System

Sporadic reports appear of lithium effects on the heart and vascular
system, though these have not been widely studied. The most common
effect seen is depression or inversion of the T wave of the electro-
cardiogram (ECG) [189], though occasional reports of transmission defects occur [190]. One such report was of a refractory cardiac ar-
rhythmia which was reversed by administration of magnesium [191]. A
detailed discussion of the effects of lithium on the ECG lists 10
reports from the literature [192] and also 4 reports of mycardiopathy.

One of the most disturbing findings, however, is that in the
infants born to women who received lithium at some stage during

pregnancy, a very high proportion of cardiovascular defects has been seen [193,194]. Of 225 babies reported to the "register of lithium babies," 11% were reported to be malformed and 8% (that is, 72% of all malformed babies) had Ebstein's anomaly, which is a major malformation of the heart. It is considered that the total number of babies born from mothers who received lithium at some stage is probably underreported and therefore the percentage of malformed infants is exaggerated by the "lithium register." However, the risk of abnormality must be considered to be high and the case for starting lithium treatment must be correspondingly strong. Furthermore, the patient should be warned of the risk so that she may be encouraged to use efficient contraception and discontinue lithium therapy some months before attempting to become pregnant. On the other hand, the number of defects reported is small and must be weighed against the dangers to the mother of not continuing lithium treatment for manic-depressive psychoses.

4.4.6. *Miscellaneous Effects*

A number of side effects may bear a relationship to the mode of action. It has been reported that weight gain is seen in a number of patients, though it has been suggested that this is a result of the better mental and physical state following lithium treatment. A number of factors already identified are involved in the regulation of body weight--for instance, carbohydrate metabolism, fluid and electrolyte balance, and feeding behavior [195].

Changes have been reported in the excretion of alkaline earth metals and phosphate following lithium [196] and a common finding is that plasma magnesium is elevated following the initiation of lithium therapy (see Ref. 197). This was not found after long-term treatment and the suggestion was made that the effects previously reported were an acute response to the initiation of therapy [197,198]. However, there have been reports of both increased and decreased calcium excretion, and the secretion of parathormone was reported to be elevated [199]. Others have suggested that the renal response to parathormone is decreased with lithium [200].

5. CONCLUSIONS

In the present review I have attempted to highlight some of the current areas of research activity in the pharmacology of lithium and also to provide a basis for the understanding of the way in which the drug is used in psychiatry today. It is clear that this review often reflects my own personal viewpoint, but I hope that sufficient reference has been made to allow the reader ready access to alternative interpretations. Lithium is of immense benefit to a large number of patients who are now living once again in the community. It is hoped that the interdisciplinary approach to research on lithium may lead ultimately to a better understanding of the mode of action and hence to more rational therapy.

REFERENCES

1. R. P. Hullin, in *Lithium in Medical Practice* (F. N. Johnson and S. Johnson, eds.), M.T.P. Press, Lancaster, U.K., 1978.

2. S. Gershon and B. Shopsin, eds., *Lithium: Its Role in Psychiatric Research and Treatment,* Plenum Press, New York, 1973.

3. F. N. Johnson, ed., *Lithium Research and Therapy,* Academic Press, London, 1975.

4. F. N. Johnson and S. Johnson, eds., *Lithium in Medical Practice,* M.T.P. Press, Lancaster, U.K., 1978.

5. F. N. Johnson, ed., *Handbook of Lithium Therapy,* M.T.P. Press, Lancaster, U.K., 1980.

6. W. E. Bunney and D. L. Murphy, eds., *The Neurobiology of Lithium,* Neurosciences Research Program Bulletin No. 14, Boston, 1976, p. 111.

7. T. B. Cooper, S. Gershon, N. S. Kline, and M. Schou, eds., *Lithium: Controversies and Unresolved Issues,* International Congress Series No. 478, Excerpta Medica, Amsterdam, 1979.

8. A. H. Rossof and W. A. Robinson, eds., *Lithium Effects on Granulopoiesis and Immune Function,* Advances in Experimental Biology No. 127, Plenum Press, New York, 1980.

9. M. Schou, *Psychopharmacol. Bull.,* 5(4), 33 (1969).

10. M. Schou, *Psychopharmacol. Bull.,* 8(4), 36 (1972).

11. M. Schou, *Psychopharmacol. Bull.*, *12*(1), 49; *12*(2), 69; *12*(3), 86 (1976).

12. M. Schou, *Neuropsychobiology, 2,* 161 (1976).

13. M. Schou, *Neuropsychobiology, 4,* 40-64 (1978).

14. M. Schou, *Neuropsychobiology, 5,* 241-265 (1979).

15. M. Schou, *Neuropsychobiology, 6,* 1-28 (1980).

16. M. Schou, *Neuropsychobiology,* in press.

17. A. Arfwedson, *Fysik, Kemi, Mineral, 6,* 145 (1818).

18. A. B. Garrod, *Gout and Rheumatic Gout,* Walton and Maberly, London, 1859.

19. G. Thilenius, *Handbuch der Balneotherapie,* Hirschwald, Berlin, 1882.

20. P. W. Squire, *Companion to British Pharmacopoeia,* 19th ed., Churchill, London, 1916.

21. A. L. Daniels, *Arch. Int. Med., 13,* 480 (1914).

22. S. A. Cleaveland, *J. Amer. Med. Ass., 60,* 722 (1913).

23. J. Biddle, *Materia Medica and Therapeutics for Physicians and Students,* 12th ed., Blakiston's, Philadelphia, 1892.

24. J. H. Talbott, *Arch. Int. Med., 85,* 1 (1950).

25. L. W. Hanlon, M. Romaine, F. J. Gilroy, and J. E. Dietrick, *J. Amer. Med. Ass., 139,* 688 (1949).

26. J. F. J. Cade, *Med. J. Aust., 36,* 349 (1949).

27. J. F. J. Cade, in *Lithium Research and Therapy* (F. N. Johnson, ed.), Academic Press, London, 1978, p. 5.

28. A. C. Corcoran, R. D. Taylor, and I. H. Page, *J. Amer. Med. Ass., 139,* 685 (1949).

29. M. Schou, *Brit. J. Psychiat., 135,* 97 (1979).

30. F. A. Jenner, L. R. Gjessing, J. R. Cox, A. Davies-Jones, R. P. Hullin, and S. M. Hanna, *Brit. J. Psychiat., 113,* 895 (1967).

31. S. M. Hanna, F. A. Jenner, I. B. Pearson, G. A. Sampson, and E. A. Thompson, *Brit. J. Psychiat., 121,* 271 (1972).

32. B. Hess and B. Chance, in *Theoretical Chemistry: Periodicities in Chemistry and Biology* (H. Eyring and D. Henderson, eds.), Academic Press, New York, 1978, p. 159.

33. W. E. Bunney, D. L. Murphy, F. K. Goodwin, and G. F. Borge, *Arch. Gen. Psychiat., 27,* 295, 304, 312 (1972).

34. M. N. E. Allsopp, M. J. Levell, S. R. Stitch, and R. P. Hullin, *Brit. J. Psychiat., 120,* 399 (1972).

35. M. Schou and K. Thomsen, in *Lithium Research and Therapy* (F. N. Johnson, ed.), Academic Press, London, 1975, p. 63.

36. P. C. Baastrup, in *Handbook of Lithium Therapy* (F. N. Johnson, ed.), M.T.P. Press, Lancaster, U.K., 1980, p. 26.

37. J. Mendels, T. A. Ramsey, W. L. Dyson, and A. Frazer, in *Lithium: Controversies and Unresolved Issues* (T. B. Cooper, S. Gershon, N. S. Kline, and M. Schou, eds.), International Congress Series No. 478, Excerpta Medica, Amsterdam, 1979, p. 35.

38. M. Schou, in *Lithium in Medical Practice* (F. N. Johnson and S. Johnson, eds.), M.T.P. Press, Lancaster, U.K., 1978, p. 21.

39. M. Schou, in *Handbook of Lithium Therapy* (F. N. Johnson, ed.), M.T.P. Press, Lancaster, U.K., 1980, p. 68.

40. M. Schou, in *Handbook of Lithium Therapy* (F. N. Johnson, ed.), M.T.P. Press, Lancaster, U.K., 1980, p. 73.

41. T. C. Jerram and R. McDonald, in *Lithium in Medical Practice* (F. N. Johnson and S. Johnson, eds.), M.T.P. Press, Lancaster, U.K., 1978, p. 407.

42. R. P. Hullin, in *Handbook of Lithium Therapy* (F. N. Johnson, ed.), M.T.P. Press, Lancaster, U.K., 1980, p. 243.

43. N. J. Birch and R. P. Hullin, *Brit. Med. J., 280,* 1148 (1980).

44. D. S. Hewick, P. Newbury, S. Hopwood, G. Naylor, and J. Moody, *Brit. J. Clin. Pharmacol., 4,* 201 (1977).

45. N. J. Birch, A. A. Greenfield, and R. P. Hullin, *Int. Pharmacopsychiat., 15,* 91 (1980).

46. M. Schou, *Psychiat. Neurol. Neurochir.* (Amsterdam), *76,* 511 (1973).

47. T. B. Cooper, P.-E. E. Bergner, and G. M. Simpson, *Amer. J. Psychiat., 130,* 601 (1973).

48. T. B. Cooper and G. M. Simpson, *Amer. J. Psychiat., 133,* 440 (1976).

49. G. M. Dempsey and H. L. Meltzer, in *Neurotoxicology* (L. Roizin, H. Shiraki, and N. Grcevic, eds.), Raven Press, New York, 1977, p. 171.

50. K. Thomsen, *J. Pharmacol. Exp. Ther., 199,* 483 (1976).

51. K. Thomsen, in *Lithium: Controversies and Unresolved Issues* (T. B. Cooper, S. Gershon, N. S. Kline, and M. Schou, eds.), International Congress Series No. 478, Excerpta Medica, Amsterdam, 1979, p. 619.

52. K. Thomsen and O. V. Olesen, *Gen. Pharmacol., 9,* 85 (1978).

53. R. J. Kerry, in *Lithium in Medical Practice* (F. N. Johnson and S. Johnson, eds.), M.T.P. Press, Lancaster, U.K., 1978, p. 337.

54. R. J. P. Williams, in *Lithium: Its Role in Psychiatric Research and Treatment* (S. Gershon and B. Shopsin, eds.), Plenum Press, New York, 1973, p. 15.

55. R. J. P. Williams, in *The Neurobiology of Lithium* (W. E. Bunney and D. L. Murphy, eds.), Neurosciences Research Program Bulletin No. 14, Boston, 1976, p. 145.

56. K. H. Stern and E. S. Amis, *Chem. Rev., 59,* 1 (1959).

57. C. Moore and B. C. Pressman, *Biochem. Biophys. Res. Commun., 15,* 562 (1964).

58. D. E. Fenton, *Chem. Soc. Rev., 6,* 325 (1977).

59. J.-M. Lehn, in *The Neurobiology of Lithium* (W. E. Bunney and D. L. Murphy, eds.), Neurosciences Research Program Bulletin No. 14, Boston, 1976, p. 133.

60. J. J. R. Frausto da Silva and R. J. P. Williams, *Nature, 263,* 237 (1976).

61. V. D. Davenport, *Amer. J. Physiol., 163,* 633 (1950).

62. N. J. Birch, in *Lithium in Medical Practice* (F. N. Johnson and S. Johnson, eds.), M.T.P. Press, Lancaster, U.K., 1978, p. 89.

63. N. J. Birch, in *New Trends in Bio-organic Chemistry* (R. J. P. Williams and J. J. R. Frausto da Silva, eds.), Academic Press, London, 1978, p. 389.

64. M. Schou, *Ann. Rev. Pharmacol., 16,* 231 (1976).

65. J. M. Davis and W. E. Fann, *Ann. Rev. Pharmacol., 11,* 285 (1971).

66. M. Schou, *Pharmacol. Rev., 9,* 17 (1957).

67. O. V. Olesen, M. Schou, and K. Thomsen, *Neuropsychobiology, 2,* 134 (1976).

68. K. Thomsen, O. V. Olesen, J. Jensen, and M. Schou, in *Current Developments in Psychopharmacology,* Vol. 3 (L. Valzelli and W. B. Essman, eds.), Spectrum, New York, 1976, p. 157.

69. M. Schou, in *Handbook of Lithium Therapy* (F. N. Johnson, ed.), M.T.P. Press, Lancaster, U.K., 1980, p. 237.

70. S. P. Tyrer, R. P. Hullin, N. J. Birch, and J. C. Goodwin, *Psychological Medicine, 6,* 51 (1976).

71. R. P. Hullin, in *Lithium: Controversies and Unresolved Issues* (T. B. Cooper, S. Gershon, N. S. Kline, and M. Schou, eds.), International Congress Series No. 478, Excerpta Medica, Amsterdam, 1979, p. 341.

72. F. N. Johnson, in *Handbook of Lithium Therapy* (F. N. Johnson, ed.), M.T.P. Press, Lancaster, U.K., 1980, pp. 219, 225.

73. N. J. Birch, *Clin. Sci. Mol. Med., 46,* 409 (1974).

74. S. C. Berens and J. Wolff, in *Lithium Research and Therapy* (F. N. Johnson, ed.), Academic Press, London, 1975, p. 443.

75. A. Desgrez and J. Meunier, *C. R. Acad. Sci., Paris, 185,* 160 (1927).

76. M. S. Ebadi, V. J. Simmonds, M. J. Hendrickson, and P. S. Lacy, *Eur. J. Pharmacol., 27,* 324 (1974).

77. S. Stern, A. Frazer, J. Mendels, and C. Frustaci, *Life Sci., 20,* 1669 (1977).

78. N. J. Birch and R. P. Hullin, *Life Sci., 11*(2), 1095 (1972).

79. N. J. Birch and F. A. Jenner, *Brit. J. Pharmacol., 47,* 586 (1973).

80. E. L. Patt, E. E. Piulett, and B. L. O'Dell, *Bioinorgan. Chem., 9,* 299 (1978).

81. N. J. Birch, D. Robinson, R. A. Inie, and R. P. Hullin, *J. Pharm. Pharmacol., 30,* 683 (1978).

82. P. A. Bond, B. A. Brooks, and A. Judd, *Brit. J. Pharmacol., 53,* 235 (1975).

83. S. Edelfors, *Acta Pharmacol. Toxicol., 37,* 387 (1975).

84. B. P. Mukherjee, P. J. Bailey, and S. N. Pradhan, *Psychopharmacology, 48,* 119 (1978).

85. M. Thellier, T. Stelz, and J.-C. Wissocq, *J. Microscopic Biol. Cell, 27,* 157 (1976).

86. J.-C. Wissocq, C. Heurteux, J. C. Bisconte, and M. Thellier, *J. Histochem. Cytochem., 27,* 1462 (1979).

87. M. Thellier, J.-C. Wissocq, and C. Heurteux, *Nature, 283,* 299 (1980).

88. H. Welfle, B. Henkel, and H. Bielka, *Acta Biol. Med. Ger., 35,* 401 (1976).

89. H. Meisner, *Biochemistry, 10,* 3485 (1971).

90. A. R. Krall, *Life. Sci., 6,* 1339 (1967).

91. M. Crompton, M. Capano, and E. Carafoli, *Eur. J. Biochem., 69,* 453 (1976).

92. B. Bhattacharya and J. Wolff, *Biochem. Biophys. Res. Commun., 73,* 383 (1976).

93. D. M. Shaw, in *Lithium Research and Therapy* (F. N. Johnson, ed.), Academic Press, London, 1975, p. 411; and in *Lithium in Medical Practice* (F. N. Johnson and S. Johnson, eds.), M.T.P. Press, Lancaster, U.K., 1978, p. 115.

94. W. E. Bunney, A. Pert, J. Rosenblatt, C. B. Pert, and D. Gallaper, *Arch. Gen. Psychiat., 36,* 898 (1979).

95. A. J. Mandell and S. Knapp, in *Lithium: Controversies and Unresolved Issues* (T. B. Cooper, S. Gershon, N. S. Kline, and M. Schou, eds.), International Congress Series No. 478, Excerpta Medica, Amsterdam, 1979, p. 789.

308   BIRCH

96. J. M. Davis, R. Coburn, D. Murphy, and D. S. Robinson, in
     Lithium: Controversies and Unresolved Issues (T. B. Cooper,
     S. Gershon, N. S. Kline, and M. Schou, eds.), International
     Congress Series No. 478, Excerpta Medica, Amsterdam, 1979, p.
     834.

97. D. L. Murphy and J. L. Costa, in *Lithium: Controversies and
     Unresolved Issues* (T. B. Cooper, S. Gershon, N. S. Kline, and
     M. Schou, eds.), International Congress Series No. 478,
     Excerpta Medica, Amsterdam, 1979, p. 815.

98. D. L. Murphy and A. J. Mandell, in *The Neurobiology of Lithium*
     (W. E. Bunney and D. L. Murphy, eds.), Neurosciences Research
     Program Bulletin No. 14, Boston, 1976, p. 165.

99. G. Balan, D. Cernatescu, M. Zrandafirescu, and L. Ababei, *Rev.
     Med.-Chir. Soc. Med. Iasi., 78,* 901 (1974).

100. G. W. Grier, L. C. Davis, and W. D. Pfeifer, *Horm. Met. Res.,
     8,* 379 (1976).

101. P. K. Kajda and N. J. Birch, *J. Inorgan Biochem,* in press.

102. K. Nakashima and S. Tuboi, *J. Biol. Chem., 251,* 4315 (1976).

103. P. K. Kajda and N. J. Birch, unpublished observations (1981).

104. N. S. Agar, M. A. Gruca, J. D. Gupta, and J. D. Harley,
     Lancet, i, 1040 (1975).

105. P. K. Kajda, N. J. Birch, M. J. O'Brien, and R. P. Hullin, *J.
     Inorgan. Biochem., 11,* 361 (1979).

106. J. A. Rillema and R. D. Smith, *Proc. Soc. Exp. Biol. Med.,
     149,* 573 (1975).

107. L. H. Lazarus and N. Kitron, *Lancet, ii,* 225 (1974).

108. B. Kadis, *Bioinorg. Chem., 6,* 183 (1976).

109. L. A. Abreu and R. R. Abreu, *Experientia, 29,* 446 (1973).

110. L. A. Abreu and R. R. Abreu, *Nature New Biol., 236,* 254 (1972).

111. H. Bera and G. C. Chatterjee, *Biochem. Pharmacol., 25,* 1554
     (1976).

112. E. S. Vizi, in *Lithium Research and Therapy* (F. N. Johnson,
     ed.), Academic Press, London, 1975, p. 391.

113. T. Tobin, T. Akera, S. S. Han, and T. M. Brody, *Mol. Pharma-
     col., 10,* 501 (1974).

114. H. W. Reading, A. J. Dewar, and N. Kinloch, *Biochem. Soc.
     Trans., 2,* 507 (1974).

115. S. J. Choi and M. A. Taylor, *Lancet, ii,* 1080 (1976).

116. N. J. Birch and I. Goulding, *Analyt. Biochem., 66,* 293 (1975).

117. A. S. Mildvan, in *Advances in Enzymology,* Vol. 49 (A. Meister,
     ed.), Wiley, New York, 1979, p. 103.

118. N. J. Birch, G. Foster, and P. K. Kajda, *Brit. J. Pharmacol.*, *72*, 506P (1981).

119. J. W. Akitt and N. J. Birch, unpublished observations (1980).

120. N. J. Birch, P. K. Kajda, and R. Wareing, *Brit. J. Pharmacol.*, *72*, 567P (1981).

121. E. T. Mellerup, P. Plenge, and O. J. Rafaelsen, *Dan. Med. Bull.*, *21*, 88 (1974).

122. E. T. Mellerup and O. J. Rafaelsen, in *Lithium Research and Therapy* (F. N. Johnson, ed.), Academic Press, London, 1975, p. 381.

123. P. B. Vendsborg and O. J. Rafaelsen, *Acta Psychiat. Scand.*, *49*, 601 (1973).

124. E. S. Haugaard, A. Frazer, J. Mendels, and N. Haugaard, *Biochem. Pharamacol.*, *24*, 1187 (1975).

125. H. Rasmussen, D. P. Goodman, N. Friedman, J. E. Allen, and K. Kurokawa, in *Handbook of Physiology*, Sec. 7, Vol. 7, American Physiology Society, Washington, D.C., 1976, p. 225.

126. R. P. Ebstein and R. H. Belmaker, in *Lithium: Controversies and Unresolved Issues* (T. B. Cooper, S. Gershon, N. S. Kline, and M. Schou, eds.), International Congress Series No. 478, Excerpta Medica, Amsterdam, 1979, p. 703.

127. J. Forn, in *Lithium Research and Therapy* (F. N. Johnson, ed.), Academic Press, London, 1975, p. 485.

128. E. Friedman, in *Lithium: Its Role in Psychiatric Research and Treatment* (S. Gershon and B. Shopsin, eds.), Plenum Press, New York, 1973, p. 75.

129. A. Geisler, P. Bech, M. Johanneson, and O. J. Rafaelsen, *Neuropsychobiology*, *2*, 211 (1976).

130. S. Christensen and A. Geisler, *Acta Pharmacol. Toxicol.*, *40*, 447 (1977).

131. P. B. Schimmer, *Biochim. Biophys. Acta*, *252*, 567 (1971).

132. P. B. Schimmer, *Biochim. Biophys. Acta*, *327*, 186 (1973).

133. D. L. Murphy, C. Donnelly, and J. Moskowitz, *Clin. Pharmacol. Ther.*, *14*, 810 (1873).

134. Y. C. Wang, G. N. Pandey, J. Mendels, and A. Frazer, *Biochem. Pharmacol.*, *22*, 845 (1974).

135. A. M. Spiegel, R. H. Gerner, D. L. Murphy, and G. D. Aurbach, *J. Clin. Endocrinol. Metab.*, *43*, 1390 (1976).

136. T. H. Steele, *J. Pharmacol. Exp. Ther.*, *197*, 206 (1976).

137. M. Zatz, *J. Neurochem.*, *32*, 1315 (1979).

138. A. Frazer, J. Mendels, S. K. Secunda, C. M. Cochrane, and C. P. Bianchi, *J. Psychiat. Res.*, *10*, 1, 1973.

139. A. Elizur, B. Shopsin, S. Gershon, and A. Ehlenberger, *Clin. Pharmacol. Ther., 13,* 947 (1972).

140. J. Mendels and A. Frazer, *J. Psychiat. Res., 10,* 9 (1973).

141. C. R. Lee, S. M. Hill, M. Dimitrakoudi, F. A. Jenner, and R. J. Pollitt, *Brit. J. Psychiat., 127,* 596 (1975).

142. J. Rybakowski, M. Chlopocka, Z. Kapelski, B. Hernacka, Z. Szajnerman, and K. Kasprzak, *Int. Pharmacopsychiat., 9,* 166 (1974).

143. J. Duhm and B. Becker, *J. Memb. Biol., 51,* 263 (1979).

144. W. Greil and F. Eisenried, in *Lithium in Medical Practice* (F. N. Johnson and S. Johnson, eds.), M.T.P. Press, Lancaster, U.K., 1978, p. 415.

145. G. N. Pandey, E. Dorus, J. M. Davis, and D. C. Tosteson, *Arch. Gen. Psychiat., 36,* 902 (1979).

146. E. Dorus, G. N. Pandey, R. Shaughnessy, M. Gavaria, E. Val, S. Ericksen, and J. M. Davis, *Science, 205,* 932 (1979).

147. H. L. Meltzer, C. J. Rosoff, S. Kassir, and R. R. Fieve, *Life Sci., 19,* 371 (1976).

148. A. Frazer, J. Mendels, and D. Brunswick, *Commun. Psychopharmacol., 1,* 255 (1977).

149. H. L. Meltzer, S. Kassir, D. L. Dunner, and R. R. Fieve, *Psychopharmacology, 54,* 113 (1977).

150. K. Martin, in *Lithium in Medical Practice* (F. N. Johnson and S. Johnson, eds.), M.T.P. Press, Lancaster, U.K., 1978, p. 167.

151. C. Lingsch and K. Martin, *Brit. J. Pharmacol., 57,* 323 (1976).

152. M. Canessa, N. Adragna, H. S. Solomon, T. M. Connolly, and D. C. Tosteson, *N. Eng. J. Med., 302,* 772 (1980).

153. G. V. R. Born, G. Grignanai, and K. Martin, *Brit. J. Clin. Pharmacol., 9,* 321 (1980).

154. A. Coppen, C. Swade, and K. Wood, *Brit. J. Psychiat., 136,* 235 (1980).

155. A. I. M. Glen, M. W. B. Bradbury, and J. Wilson, *Nature, 239,* 399 (1972).

156. L. A. Beaugé and E. del Campillo, *Biochim. Biophys. Acta, 433,* 547 (1976).

157. A. I. M. Glen, in *Lithium: Controversies and Unresolved Issues* (T. B. Cooper, S. Gershon, N. S. Kline, and M. Schou, eds.), International Congress Series No. 478, Excerpta Medica, Amsterdam, 1979, p. 768.

158. H. W. Reading and T. Isbir, *Biochem. Pharmacol., 28,* 3471 (1979).

159. A. I. M. Glen, in *Lithium in Medical Practice* (F. N. Johnson and S. Johnson, eds.), M.T.P. Press, Lancaster, U.K., 1978, p. 183.

160. R. P. Hullin and N. J. Birch, in *Lithium: Controversies and Unresolved Issues* (T. B. Cooper, S. Gershon, N. S. Kline, and M. Schou, eds.), International Congress Series No. 478, Excerpta Medica, Amsterdam, 1979, p. 584.

161. R. P. Hullin, in *Lithium in Medical Practice* (F. N. Johnson and S. Johnson, eds.), M.T.P. Press, Lancaster, U.K., 1978, p. 433.

162. J. Wolff, in *The Neurobiology of Lithium* (W. E. Bunney and D. L. Murphy, eds.), Neuroscience Research Program Bulletin No. 14, Boston, 1976, p. 178.

163. P. J. Mannisto, J. Leppaluoto, and P. Virkkunen, *Acta Endocrinol.*, *74*, 492 (1973).

164. S. C. Berens, R. S. Bernstein, J. Robbins, and J. Wolff, *J. Clin. Invest.*, *49*, 1357 (1970).

165. J. H. Lazarus and E. H. Bennie, *Acta Endocrinol.*, *70*, 226 (1972).

166. C. Christiansen, P. C. Baastrup, and I. Transbøl, *Lancet*, *ii*, 969 (1976).

167. P. C. Baastrup, C. Christiansen, and I. Transbøl, *Acta Psychiat. Scand.*, *57*, 124 (1978).

168. N. J. Birch, in *Handbook of Lithium Therapy* (F. N. Johnson, ed.), M.T.P. Press, Lancaster, U.K., 1980, p. 365.

169. J. Hestbech, H. E. Hansen, A. Amdisen, and S. Olsen, *Kidney Int.*, *12*, 205 (1977).

170. H. E. Hansen, J. Hestbech, J. L. Sorensen, K. Norgaard, J. Heilskov, and A. Amdisen, *Quart. J. Med.*, *48*, 577 (1979).

171. G. D. Burrows, B. Davis, and P. Kincaid-Smith, *Lancet*, *i*, 1310 (1978).

172. F. A. Jenner and P. R. Eastwood, in *Lithium in Medical Practice* (F. N. Johnson and S. Johnson, eds.), M.T.P. Press, Lancaster, U.K., 1978, p. 247.

173. J. H. Holmes, in *Thirst* (M. J. Wayner, ed.), Pergamon Press, Oxford, 1964, p. 57.

174. N. J. Birch and R. P. Hullin, *Brit. Med. J.*, *280*, 1148 (1980).

175. F. A. Jenner, *Arch. Gen. Psychiat.*, *36*, 888 (1979).

176. Editorial, *Lancet*, *ii*, 1056 (1979).

177. P. Vestergaard, in *Handbook of Lithium Therapy* (F. N. Johnson, ed.), M.T.P. Press, Lancaster, U.K., 1980, p. 345.

178. H. E. Hansen and A. Amdisen, *Quart. J. Med.*, *47*, 123 (1978).

179. R. P. Hullin, V. P. Coley, N. J. Birch, D. B. Morgan, and T. H. Thomas, *Brit. Med. J.*, *1*, 1457 (1979).

180. A. I. M. Glen, M. Dodd, E. B. Hulme, and N. Kreitman, *Neuropsychobiology*, *5*, 167 (1979).

181. S. Christensen and B. Hansen, *Renal Physiol.*, in press.

182. I. Singer, in *The Neurobiology of Lithium* (W. E. Bunney and D. L. Murphy, eds.), Neurosciences Research Program Bulletin No. 14, Boston, 1976, p. 175.

183. M. Cox and I. Singer, in *Lithium: Controversies and Unresolved Issues* (T. B. Cooper, S. Gershon, N. S. Kline, and M. Schou, eds.), International Congress Series No. 478, Excerpta Medica, Amsterdam, 1979, p. 646.

184. I. Singer and M. Cox, in *Lithium: Controversies and Unresolved Issues* (T. B. Cooper, S. Gershon, N. S. Kline, and M. Schou, eds.), International Congress Series No. 478, Excerpta Medica, Amsterdam, 1979, p. 642.

185. T. H. Steele, B. A. Stromberg, and J. L. Underwood, in *Lithium: Controversies and Unresolved Issues* (T. B. Cooper, S. Gershon, N. S. Kline, and M. Schou, eds.), International Congress Series No. 478, Excerpta Medica, Amsterdam, 1979, p. 656.

186. M. Dimitrakoudi and F. A. Jenner, in *Lithium Research and Therapy* (F. N. Johnson, ed.), Academic Press, London, 1975, p. 507.

187. S. P. Tyrer and B. Shopsin, in *Handbook of Lithium Therapy* (F. N. Johnson, ed.), M.T.P. Press, Lancaster, U.K., 1980, p. 289.

188. W. Girke, F.-A. Krebs, and B. Muller-Oerlinghausen, *Pharmakopsychiatry*, *10*, 24 (1975).

189. R. G. Demers and G. Heninger, *Dis. Nerv. Syst.*, *31*, 674 (1970).

190. C. M. Jaffe, *Amer. J. Psychiat.*, *134*, 88 (1977).

191. L. I. G. Worthley, *Anaesth. Intens. Care*, *2*, 357 (1974).

192. J. W. Albrecht and B. Muller-Oerlinghausen, in *Handbook of Lithium Therapy* (F. N. Johnson, ed.), M.T.P. Press, Lancaster, U.K., 1980, p. 323.

193. M. Weinstein, in *Handbook of Lithium Therapy* (F. N. Johnson, ed.), M.T.P. Press, Lancaster, U.K., 1980, p. 421.

194. M. Weinstein, in *Lithium: Controversies and Unresolved Issues* (T. B. Cooper, S. Gershon, N. S. Kline, and M. Schou, eds.), International Congress Series No. 478, Excerpta Medica, Amsterdam, 1979, p. 432.

195. N. J. Birch, *Trends Biochem. Sci.*, *2*, 282 (1977).

196. E. T. Mellerup, B. Lauritsen, H. Dam, and O. J. Rafaelsen, *Acta Psychiat. Scand.*, *53*, 360 (1976).

197. N. J. Birch, A. A. Greenfield, and R. P. Hullin, *Psychological Medicine, 7,* 613 (1977).

198. D. L. Dunner, H. L. Meltzer, H. C. Schreiner, and J. L. Fiegelson, *Acta Psychiat. Scand., 51,* 104 (1975).

199. C. Christiansen, P. C. Baastrup, P. Lindgren, and I. Transbøl, *Acta Endocrinol., 88,* 528 (1978).

200. R. H. Gerner, R. M. Post, A. M. Spiegel, and D. L. Murphy, *Biol. Psychiat., 12,* 145 (1977).

AUTHOR INDEX

Numbers in parentheses are reference numbers and indicate that an author's work is referred to although his name may not be cited in the text. Underlined numbers give the page on which the complete reference is listed.

Printed and bound by CPI Group (UK) Ltd, Croydon, CR0 4YY

23/10/2024

01778224-0001